Handbook of Scientific Proposal Writing

Handbook of Scientific Proposal Writing

A. Yavuz Oruç

University of Maryland
College Park, USA

CRC Press
Taylor & Francis Group
Boca Raton   London   New York

CRC Press is an imprint of the
Taylor & Francis Group, an **informa** business

A CHAPMAN & HALL BOOK

CRC Press
Taylor & Francis Group
6000 Broken Sound Parkway NW, Suite 300
Boca Raton, FL 33487-2742

First issued in paperback 2017

© 2012 by Taylor & Francis Group, LLC
CRC Press is an imprint of Taylor & Francis Group, an Informa business

No claim to original U.S. Government works

ISBN-13: 978-1-4398-6918-5 (hbk)
ISBN-13: 978-1-138-11420-3 (pbk)

Library of Congress Cataloging-in-Publication Data

Oruc, A. Yavuz.
  Handbook of scientific proposal writing / author, A. Yavuz Oruc.
    p. cm.
  Includes bibliographical references and index.
  ISBN 978-1-4398-6918-5 (hardcover : alk. paper)
  1. Proposal writing in research--Handbooks, manuals, etc. I. Title.

Q180.55.P7O78 2012
808.066'5--dc23   2011036560

Visit the Taylor & Francis Web site at
http://www.taylorandfrancis.com

and the CRC Press Web site at
http://www.crcpress.com

Dedication

To my friends who inspired me to write this book

Contents

Preface

This book is written primarily to offer researchers and research administrators in academia and funding agencies a broad perspective on the process of initiating and conducting funded scientific research projects. It is also written to provide researchers with a hands-on approach for conceiving initial research ideas, expanding them to competitive proposals, and transforming them to concrete new research results.

Researchers in academia and other scientific institutions and laboratories routinely engage in proposal writing activities with great desire and optimism to win awards and carry out their research projects. This sense of optimism is frequently replaced by frustration and disillusionment when they hear that their proposals are declined after a considerable period of waiting. At the other end of the funding process, research administrators need to keep up with changing funding priorities of their agencies and struggle to maintain a steady flow of proposals to their programs to make their awards more competitive. Seminal and inventive research ideas incubate in this trial-and-error process of submitting and evaluating research proposals but writing a competitive proposal requires a lot more than having such ideas.

Understanding that submitting a proposal to a funding agency is only one dimension of a multivariable and complex funding process is a good first step toward unlocking the puzzle behind why some research proposals receive awards and others are declined. For one, funding agencies, just like any other organization, have missions to accomplish, and their missions are codified in the funding operations they conduct. For example, a funding agency that is focused on the development of certain technologies would be more interested in goal-driven research projects rather than supporting researchers to carry out projects of their own interests. In contrast, funding agencies such as the U.S. National Science Foundation (NSF) and European Research Council (ERC) fund basic research projects that are subject to peer review. To win NSF or ERC awards, researchers must convince their colleagues in their fields of activity that (1) they have original research ideas that are worth pursuing; and (2) they have the talent, skills, and resources to pursue them. Thus, understanding the mission of a funding agency is crucial to writing a competitive proposal. A number of other factors play an equally critical role in winning research grants and awards. Most funding agencies expect research proposals to possess three interrelated qualities: intellectual merit, broader impact, and feasibility or likelihood of success. Proposals that fail to demonstrate any one of these three criteria would be a hard sell to any funding agency. Those that succeed in all three respects may still face stiff competition, but they would at least fall within a small pool of competitive research proposals. It should be emphasized that home institutions of researchers also play a significant role in drawing funds to their research projects. The condition of research facilities and laboratories significantly impacts the research capability of investigators, especially in applied and experiment research fields. Universities and other research institutions can do much to boost the chances of their investigators to write competitive proposals and receive awards by building and upgrading their research infrastructure and facilities. They can also help create an environment that promotes scientific research in all of its manifestations and not only to raise funds. The dynamics of proposal submission and evaluation process is crucial to winning research grants as well. Investigators conduct their research projects in a field of activity where other researchers work to compete with them. Researchers' level of activity and impact of their previous research results have a significant effect on their success rates of winning research grants. Last but not least, participation in the peer-review evaluations of the research proposals of fellow researchers in the field is essential to demonstrating that a researcher is active in a field and has a say in determining which proposals receive funding. It also provides access to incubating research ideas and helps master the process of writing and submitting competitive proposals.

The book begins in Chapter 1 with a survey of key scientific discoveries to date, and an overview of scientific research and its methods. It describes the interplay between basic and applied research and explains how they drive technological innovations. In Chapter 2, the intellectual merit and broader impact criteria are described, and an analysis of what makes a scientific study widely recognized is provided. A case study of a highly cited paper in anthropology is presented to demonstrate how a high impact research work is publicized and disseminated. The critical role of having preliminary research results and their integration to existing research is explained, and engaging in collaborative research projects is shown to be an effective method to synergize preliminary results and expedite the proposal submission process. Chapter 3 focuses on how to develop preliminary ideas into a comprehensive research proposal. It describes the building blocks of a scientific research proposal and different styles of incorporating intellectual merit and broader impact into various parts of a proposal. It also discusses the potential causes of declinations of proposals and some common mistakes investigators commit during the writing and submission of proposals. Chapter 4 deals with conducting research projects. It describes how to assemble a research team and map out a plan of work for a funded research project into a work flow schedule to meet the projected deadlines. It also explains how to disseminate research results, prepare annual progress reports, and get ready for the next proposal. Chapter 5 examines the synergy between research and education. It describes how a research university can promote scientific research and motivate its faculty to build and carry out research programs. It introduces various metrics to measure the research performance of a group of faculty and implement a quantitative merit raise system. It also explains how various bibliometric indicators, such as the *h*-index, *g*-index, and impact factor, can be used to analyze and compare the performance of a group of faculty in a department or college with other similar programs in other universities. Common mistakes that universities should avoid so as not to stifle research are also discussed in this chapter. Chapter 6 focuses on the important role of funding agencies in promoting scientific research. It examines their missions and operational characteristics for facilitating the resources for research projects and ensuring an effective discharge and utilization of such resources. It introduces techniques for evaluating proposals and making funding decisions. The chapter ends with a list of potential mistakes funding agencies should avoid not to discourage investigators from submitting proposals.

The book includes more than 50 figures and more than 20 tables to help convey its key ideas. More than 150 citations appear throughout the book to provide pointers to additional sources for further reading. Examples are provided wherever needed and to help the reader ease through relatively more abstract concepts.

The book is designed as a self-contained reference for use in developing and writing scientific research proposals. The related literature includes a number of textbooks that cover proposal and grant writing topics at varying degrees of breadth and depth. Some of these are also tailored to developing proposals for specific funding agencies. The goal in this book is to offer the reader a broader perspective for capturing the essence of funded scientific research with the hope that this will lead the reader to conceiving and developing more robust proposals. To this end, Chapters 2 and 3 provide specific methods for laying out initial research ideas, expanding them into a comprehensive proposal, and timing the submission of proposals. Intellectual merit, broader impact, and preliminary research criteria described in Chapter 2 should be valuable to almost any investigator who is not sure about the timing of submitting a proposal. The proposal skeleton that is described in Chapter 3 provides a generic template that can be shaped into an effective proposal by filling it out with content. The material in Chapter 4 can help researchers to organize their research activities more effectively, especially if they are constrained in their resources and have tight deadlines to meet. The material covered in Chapter 5 should be helpful to both researchers in maximizing their research performance and administrators in utilizing their limited merit budgets more effectively and equitably. The research performance metrics introduced in this chapter can be especially useful for faculty researchers to calibrate their research accomplishments. The material in Chapter 6 is written primarily for program directors and research administrators in funding agencies. However, I believe that understanding this material thoroughly can also help investigators produce more robust proposals to overcome the potential traps that they may face during the stiff review and evaluation process at funding agencies.

This book may also be used as a text in a first-year graduate or senior-level course in scientific writing, and science and technology curricula as well as in seminar courses that focus on scientific research proposal writing and funding. The questions at the end of the chapters are designed to probe the subject deeper and might be useful in such instructional uses of the book.

Acknowledgments

This book could not be written without the generous support of many colleagues and friends. I greatly benefited from the wisdom of my former colleagues at the National Science Foundation (NSF) in shaping my thoughts early on. I owe much to the late Frank Anger, Kamal Abdali, Bob Grafton, Helen Gill, and Rita Rodrigues for teaching me what it means to be a program director at NSF. I would like to thank them and all my other former fellow program directors at the Computing and Communication Foundations Division. My association with the Scientific and Technological Research Council of Turkey (TÜBİTAK) has also been influential in authoring this book. I would like to thank my colleagues Nuket Yetiş, Abdullah Atalar, Ömer Cebeci, Ömer Anlağan, and Faruk Arınç for making me feel at home during my visits to TÜBİTAK. Thanks also to the fellow TÜBİTAK research project team members Alper Yıldırım, Cevdet Aykanat, Abdullah Atmaca, Emre Çakır, and Enver Kayaaslan for their contributions to the material covered in Chapter 6. Thanks to the University of Maryland and Bilkent University for granting me the much needed time and space to develop and complete this book project as well. Last but not least, I am grateful to Bob Stern and CRC press for finding the idea of writing this book a worthwhile endeavor and keeping me on track through all of its critical phases.

Author

A. Yavuz Oruç, Ph.D., received a B.Sc. in electrical engineering from the Middle East Technical University in 1976, a M.Sc. in electronics from the University of Wales in 1978, and a Ph.D. in electrical engineering from Syracuse University in 1983. He has conducted research in computer systems, parallel processing, and interconnection network theory, and is a recipient of a number of research grants. His more recent research has focused on quantum packet switching, an emerging field of research in quantum information processing. He was the director of the Computer Systems Architecture Program at the U.S. National Science Foundation from 2000 to 2002, and served as a senior advisor to the president of the Scientific Research and Technological Council of Turkey between 2005 and 2008.

He has been a full professor at the University of Maryland, College Park since 1995. He previously held faculty positions at Rensselaer Polytechnic Institute (RPI) and Bilkent University. His research resulted in more than 110 publications in archival journals and conference proceedings, and the supervision of 16 doctoral dissertations and 24 master's theses. He holds a patent on "System and Method for Performing Fast Algebraic Operations on a Permutation Network." He was an associate editor for the *IEEE Transactions on Parallel and Distributed Systems* between 2003 and 2007.

He is the coinventor of the CodeMill language and programming software, for which he received the 2000 Teaching with Technology Award from the University of Maryland.

chapter one

Scientific research
The fountain of progress

Scientific knowledge shifts from one frontier to the next with seminal discoveries. The seeds of these discoveries are sown in basic investigations driven by human curiosity. Nearly all scientific and technological accomplishments of today can be traced to explorations and inventions of the past. The wheels used in automobiles, buses, trains, and airplanes were invented to move carts and chariots with onagers in the ancient cities of Mesopotamia. The buoyancy principle of Archimedes is what floats boats, ships, and other vessels in water. Ubiquitous wireless technologies around us owe much to Hans Christian Orsted, Andre-Marie Ampere, James Maxwell, and Heinrich Hertz for their discovery of electromagnetism, electrodynamics, and electromagnetic and radio waves.[*] Satellites orbit the earth and spaceships travel to other planets according to the gravitational laws discovered by Isaac Newton. In the English version of *Philosophiae Naturalis Principia Mathematica*, he eloquently expressed this proliferation of scientific knowledge and wrote:[†]

> Then from these forces by other propositions, which are also mathematical, we deduce the motions of the Planets, the Comets, the Moon, and the Sea. I wish we could derive the rest of the phenomena of Nature by the same kind of reasoning from mechanical principles. For I am induced by many reasons to suspect that they may all depend on certain forces by which the particles of bodies, by some causes hitherto unknown, are either mutually impelled towards each other and cohere in regular figures, or repelled and recede from each other; which forces being unknown, researchers have hitherto attempted to search in vain. But I hope the principles here laid down will afford some light either to that, or to some truer method of Philosophy.

[*] Darrigol, O. 2000. *Electrodynamics from Ampere to Einstein*. Oxford University Press.
[†] Newton, I. 1723. *The mathematical principles of natural philosophy* (Trans. A. Motte). London.

Much happened since the monumental scientific contributions of Isaac Newton and his influential contemporaries, Galileo Galilei, Johannes Kepler, Rene Descartes, Pierre de Fermat, Blaise Pascal, Christiaan Huygens, and Gottfried Leibniz. Newton's call was answered with Werner Heisenberg's discovery of matrix mechanics and uncertainty principle* in 1925 and Erwin Schrodinger's discovery of wave mechanics and wave equation[†] in 1926. Yet, the scientific principles on which Newton built one of the most fundamental theories of the physical forces of nature remain nearly intact. It is therefore important to put the process of developing new scientific ideas and writing scientific proposals in a proper context. To that end, the aim in this chapter is (a) to provide an overview of the scientific method and progress of science to date, (b) to explain what makes an investigation of a set of problems scientific research and why some scientific studies draw more attention than others, and (c) to describe the particular aspects of scientific research in basic and applied research fields.

1.1   A brief history and origins of scientific exploration

There is strong scientific evidence that the first notable exploration of nature by humans began in Western Asia.[‡] Archaeological findings to date suggest that the scientific method has evolved from rudimentary trial-and-error techniques in agriculture, hydrology, and transportation along the banks of the Euphrates and Tigris rivers in ancient Mesopotamia. These fertile lands and rivers triggered a transition of human activity from the nomadic life of hunting and gathering to agricultural activities more than 10,000 years ago.[§,¶,**] This evidently led to settlements along the river banks in Mesopotamia beginning around 5000 B.C. Archaeological studies[††] also reveal that the settlers inscribed pictograms, ideograms, and cuneiforms on clay tablets for trading, accounting, and bookkeeping

---

* Heisenberg, W. 1925. Uber quantentheoretische Umdeutung kinematischer und mechanischer Beziehungen. *Z. Phys.* 33:879–893.

[†] Schrödinger, E. 1926. An undulatory theory of the mechanics of atoms and molecules. *Phys. Rev.* 28:1049–1070.

[‡] Postgate, J. N. 1994. *Early Mesopotamia: Society and economy at the dawn of history.* Routledge.

[§] Maisels, C. K. 1999. *The emergence of civilization from hunting and gathering to agriculture, cities, and the state in the Near East.* Routledge/Chapman & Hall.

[¶] McMahon, A. 2007. From sedentism to states, 1000–3000 B.C.E. In *A companion to the Ancient Near East,* ed. D. Snell, 20–33. Blackwell.

** Bunch, B., and Hellemans, A. 2004. *History of modern science and technology.* Houghton Mifflin Harcourt.

[††] Yushu, G. 2010. The Sumerian account of the invention of writing—A new interpretation. *Procedia Social and Behavioral Sciences* 2:7446–7453.

during early dynasties of Uruks, Sumerians, Akkadians, and Assyrians, 4400–2100 B.C. These early settlers conceived and built the first gigantic structures of worship, called ziggurats. These temples were early forms of pyramid architectures that were later built in ancient Egypt, beginning around 2700 B.C. Cuneiform texts written on tablets suggest that, by 1800 B.C., Mesopotamia was buzzing with lawyers, doctors, accountants, astronomers, teachers, artists, carpenters, and masons.[*][†] These occupations were intertwined in a complex socioeconomic structure that also included kings, priests, soldiers, farmers, potters, merchants, traders, and servants. Some of the earliest mathematical discoveries, which include the sexagesimal (base 60) number system and arithmetic, and solutions of quadratic equations, date to this period. These discoveries were carved on thousands of clay tablets using cuneiform scripts. One of these tablets, dubbed Plimpton 322, and dating to the Hammurabi dynasty in Babylon, lists 15 Pythagorean triples. This would suggest that Babylonians made use of the right-angle triangle law to compute lengths and areas long before Pythagoras established it as the theorem that carries his name.[‡] These mathematical recreations in Mesopotamia were followed by more intricate mathematical discoveries written on papyri during Egyptian dynasties, 1700 to 600 B.C. Similar discoveries have also been reported to occur in other parts of the world, in particular, during the Maya civilization in South America, Harappan civilization in Indus Valley, and ancient Chinese dynasties in Asia.[§]

With the expansion of the Achaemenid (Persian) Empire that ruled the Mesopotamian, Anatolian, and Egyptian kingdoms, and subsequent expansion of the Macedonian Empire toward the east in antiquity, the scientific discoveries of the Mesopotamian and Egyptian philosophers and mathematicians influenced the Mediterranean civilizations during the Hellenistic and Roman periods, 600 B.C. to 400 A.D. During this period, such noted philosophers, astronomers, and mathematicians as Thales, Anaximander, Pythagoras, Socrates, Hippocrates, Plato, Aristotle, Euclid, Archimedes, Eratosthenes, Apollonius, Diophantus, and Ptolemy founded and developed the conceptual and experimental facets of scientific inquiry and exploration.[¶][**]

The progress of scientific inquiry and technological explorations took a heavy toll with the fall of the Western Roman Empire and invasion of Eastern and Southern Europe by the eastern nomadic warriors, notably by

* Bunch and Hellemans, *History of modern science.*
† Macintosh, J. R. 2005. *Ancient Mesopotamia: New perspectives.* ABC-CLIO Pub.
‡ Robson, E. 2002. Words and pictures: New light on Plimpton 322. *Amer. Mathemat. Month.* 109:105–120.
§ Bunch and Hellemans, *History of modern science.*
¶ Bunch and Hellemans, *History of modern science.*
**Sedgwick, W. T., and Tyler, H.W. 1917. *A short history of science.* Macmillan.

the Huns, during the 4th and 5th centuries. At the same time, the birth of Islamic religion and philosophy in the 6th century and its rise and expansion during its golden age between the 9th and 13th centuries led to a scientific renaissance in the Eastern world.[*][†] During this golden period, polymaths like al-Khwarizmi, al-Kindi, al-Farabi, al-Biruni, Ibn Sina, Ibn al-Haytham, and Omar Khayyam made major contributions to astronomy, chemistry, mathematics, medicine, physics, and philosophy. After a millennium of relatively dark period in Europe, these scientific discoveries in the eastern world sparked the European renaissance. Trading and commerce brought precious scientific knowledge and technological knowhow from the Eastern world to the port cities on the Mediterranean around the beginning of the 14th century. Nicolaus Copernicus, Johannes Kepler, and Galileo Galilei studied and made observations of celestial bodies and planetary motions, thereby initiating a scientific revolution that parted from Aristotle's model of the universe and physical world.[‡] With Gutenberg's invention of the printing machine around 1450, these early scientific accomplishments of the European renaissance quickly spread across Europe. During the 17th century, Rene Descartes, Pierre de Fermat, Blaise Pascal, Christiaan Huygens, Isaac Newton, and Gottfried Leibniz invented the analytic geometry, and differential and integral calculus and used them to formalize and establish the laws of celestial and gravitational mechanics.[§] With the advent of calculus as a powerful tool to model and solve physical problems, several brilliant scientists, including Leonhard Euler, Joseph-Louis Lagrange, Nicholas and Daniel Bernoulli, Abraham De Moivre, Adrian Marie Legendre, Joseph Fourier, Simeon Poisson, and Pierre-Simon Laplace, made lasting contributions to mathematics and physics.[¶]

The invention of infinitesimal and differential calculus triggered an interest in automating calculations. Early explorations of John Napier and Wilhem Schickard on automating arithmetic operations were followed by Blaise Pascal's mechanical adder in 1642 and Gottfried Leibniz's stepped reckoner in 1672.[**] These early calculators led to Charles Babbage's invention of the difference and analytic engines during the first half of the 19th century.[††] Even though Babbage never completed the construction of

[*] Bunch and Hellemans, *History of modern science.*
[†] Sedgwick and Tyler, *A short history.*
[‡] Applebaum, W. 2005. *The scientific revolution and the foundations of modern science.* Greenwood Press.
[§] Applebaum, W. Ed. 2000. *Encyclopedia of the scientific revolution: From Copernicus to Newton.* Garland.
[¶] Bunch and Hellemans, *History of modern science.*
[**] Collier, B., and Maclachlan, J. 1998. *Charles Babbage and the engines of perfection.* Oxford University Press.
[††] Collier and Maclachlan, *Charles Babbage.*

either machine, he is credited with conceiving the first forerunners of the modern day computer. The design and construction of the first full-scale digital computers had to wait almost a century. Pioneering contributions of Kondrad Zuse, John Atanasoff, George Stibitz, John Mauchly, John Presper Eckert, and John von Neumann eventually led to the prototypes of modern-day computers in the 1940s.*

Within two centuries of the European renaissance, science was proliferating at new frontiers with Robert Boyle, Otto von Guericke, Antoine Lavoisier, Henry Cavendish, and Joseph Priestley laying the foundations of modern chemistry.[†] Eventually, Otto von Guericke's discovery of the air pump triggered the invention of the steam engine by Thomas Savery, Thomas Newcomen, and James Watt toward the end of the 17th and early 18th centuries.[‡] This then led to the Industrial Revolution with the mechanization of textile, manufacturing, and transportation industries in England, Europe, and North America in the second half of the 18th century.[§] These scientific breakthroughs were followed in the 19th century by the discovery of the laws of thermodynamics through the scientific quests of Sadi Carnot, William Thomson, Hermann von Helmholtz, Prescott Joule, and Ludwig Boltzmann that revealed the fundamental relations between work, heat, and energy.

Along another scientific track, William Gilbert, Charles-Augustin de Coulomb, and Benjamin Franklin were involved in the study of the behavior of electrical charge and electricity.[¶] These initial studies provided the experimental stimulus for fundamental discoveries in electromagnetism with landmark contributions by Hans Christian Oersted, André-Marie Ampère, Michael Faraday, Carl Friedrich Gauss, Gustav Kirchhoff, James Maxwell, and Heinrich Hertz in the 19th century.[**] The discoveries in physical sciences and electrical engineering then led to many remarkable inventions. These included the electric motor and generator, telegraph, telephone, light bulb, radio, television, calculators, x-ray radiography, motion picture cameras, automobiles, and electrical trains through the first quarter of the 20th century.

The laws of electromagnetic field theory ultimately led to the discovery of the electron as an elementary particle with a negative electric charge and finite mass by J. J. Thomson in 1897. During the same time period, Max Planck discovered that, in atomic scales, light does not behave like a wave as predicted by Maxwell's laws of electromagnetism and electrodynamics. Rather, it behaves like quanta of energy, called photons. This

---

* Rojas, R., and Hashagen, U. 2002. *First computers: History and architectures.* MIT Press.
[†] Asimov, I. 1965. *A short history of chemistry.* Anchor Books.
[‡] Bunch and Hellemans, *History of modern science.*
[§] Corrick, J. A. 1998. *The industrial revolution.* Lucent Books.
[¶] Bunch and Hellemans, *History of modern science.*
[**] Darrigol, *Electrodynamics from Ampere to Einstein.*

discovery led to a new model to explain the physical behavior of subatomic particles by Ernest Rutherford and Niels Bohr through a discrete orbital motion theory of electrons and atomic nucleus structure. These findings served as a spark for a new kind of physics at the turn of 20th century to explain physical phenomena at much smaller scales of physics, where electrons spin in picometer orbits around the nuclei of neutrons and protons.* These then resulted in matrix and wave mechanics models of quantum physics and set the stage for the surface and semiconductor physics that ultimately led to the invention of the transistor in 1948 by John Bardeen, Walter Brattain, and William Shockley.[†,‡] This hallmark invention opened the door for the design and development of myriad electronic gadgets, penetrating every facet of life and bringing people together around the globe with electronic cell phones, wireless laptop computers and networks of servers built with microprocessor chips, housing hundreds of millions of tiny transistors. It also led to numerous medical technologies, increasing the quality and longevity of life.

The development of electromagnetic field theory also triggered new theoretical investigations, leading to the discovery of special and general relativity laws of physics by Albert Einstein, Henry Poincare, David Hilbert, Hendrix Lorentz, and Hermann Minkowski.[§,¶] These fundamental accomplishments in physics then brought about today's nuclear energy and space technologies. More recent discoveries in material science, nanotechnology, and signal and image processing led to the design and fabrication of ultrafast integrated circuit chips and processors, acoustic and electron microscopes, computer tomography, ultrasound radiology, and other imaging devices. Advances in applied mathematics and computer science brought about the fields of computational physics, chemistry, and biology with major discoveries extending from new pharmaceutical technologies to DNA sequencing algorithms.

Needless to say, the proliferation of scientific and technological discoveries during the last 150 years has been trailblazing.[**] Every one of these major discoveries can likely be traced to a human experiment that began with three simple questions: what, why, and how. When these questions are posed within a scientific context and tackled by scientists

---

[*] Ning Yang, C. 1961. *Elementary particles: A short history of some discoveries in physics.* Princeton University Press.

[†] Bardeen, J., and Brattain, W. H. 1948. The transistor: A semiconductor triode. *Phys. Rev.* 74: 230–231.

[‡] Shockley, W. B. 1956. "Transistor technology evokes new physics," Nobel lecture. http://www.nobelprize.org.

[§] Darrigol, *Electrodynamics from Ampere to Einstein.*

[¶] Einstein, A. 1954. *Relativity: The special and general theory,* 15th ed. Routledge.

[**] For a comprehensive digest of these recent discoveries and earlier ones, the reader is referred to Bunch and Hellemans' *History of Modern Science.*

of the highest caliber, the result is often a new significant scientific discovery that can potentially be used to build new products and systems to help overcome the limitations of existing technologies. The foundation of this curiosity constitutes what is called *scientific research* as described next.

1.2 What is scientific research?

The term *scientific research* is used broadly to refer to an investigation of open problems in a scientific field. The synergy between the two words that constitute the term codifies its meaning. We will explore these two words and the synergy between them to give as complete a description of the term *scientific research* as possible. We begin with the second word first.

The *Merriam-Webster* online dictionary* defines *research* as "careful or diligent search," "studious inquiry; especially investigation or experimentation aimed at the discovery or interpretation of facts, revision of accepted theories in the light of new facts or practical application of such new or revised theories or laws," or "collecting of information about a particular subject." Others define it more succinctly as "human activity based on intellectual application in the investigation of matter." We will adopt the following more generic definition to focus on the essence of research:

Remark 1.1 Research is a process of understanding a problem, and discovering facts to help solve it.

Researching a problem can thus be viewed as solving or completing a puzzle. However, in this puzzle analogy, obtaining the missing pieces may involve steps that may go far beyond searching for them. The pieces may exist in some other form and may have to be modified before they can be used, or they may not even exist and may have to be created. Exactly how the pieces are obtained and the puzzle is solved depends on the model of research process used.

One such model is that of a mathematical investigation as shown in Figure 1.1. Mathematical results are customarily established using five basic entities: (1) axioms (postulates), (2) definitions and propositions, (3) lemmas, (4) theorems, and (5) corollaries. Axioms are generally the ground rules of a mathematical investigation. All statements must agree with the axioms of the mathematical domain within which the research activity is carried out. Definitions and propositions form the front end of such an activity. They are introduced to formalize the problems to be solved. Lemmas serve as auxiliary puzzle pieces; theorems correspond to blocks

---

* *Merriam-Webster,* http://www.merriam-webster.com/dictionary/research (accessed November 21, 2010); WebCite, http://www.webcitation.org/5uPFzDwhU.

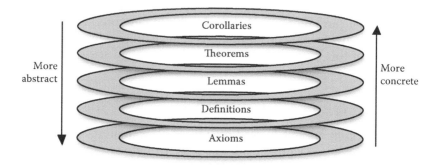

Figure 1.1 Characterization of a research activity using a mathematical model.

of puzzles, and corollaries extend the solutions of puzzles to the solutions of other similar puzzles. In this setting, stating and proving lemmas should constitute a research activity if they have not been stated or proved before. They are like creating new pieces to solve a puzzle. Stating and proving theorems is similar to putting smaller pieces together to make bigger pieces of a puzzle. This would also be a research activity if the theorems in question have not been stated and proved elsewhere. On the other hand, corollaries are more like consequences of theorems and lemmas. Stating and proving corollaries, in and of themselves, should therefore be viewed more as a process of extending known results. However, in some cases, coming up with corollaries and proving them may involve considerable creativity and effort and thus constitute a legitimate research activity.

The mathematical research model can be applied to capture the research content of projects that rely on mathematical formalisms for researching problems. For projects in physical sciences, the model shown in Figure 1.2 would be more appropriate. Just as axioms form the ground rules in a mathematical investigation, empirical or experimental observations constitute the foundation of physical research. When such observations are not accountable by existing theories and laws, new hypotheses are introduced to explain the reasons behind them. Hypotheses are then converted into models with which new laws are asserted and tested to see if they explain the empirical or experimental observations.

A third investigation model can be used to capture the style of investigations in engineering and systems research projects as shown in Figure 1.3. A conceptual blueprint represents the grand plan of such a project and defines its scope in this model. Concepts either drive new technologies or they are mapped into existing technologies. The design and simulation phase of the project is carried out to generate and develop the main building blocks of the system to be created within the chosen or defined

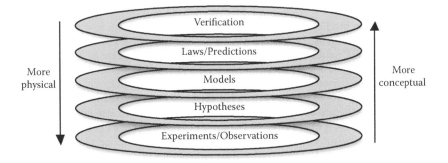

Figure 1.2 Characterization of a research activity using a physical model.

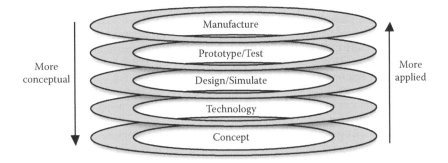

Figure 1.3 Characterization of a research activity using an engineering model.

technology base. The last two phases involve prototyping and manufacturing of the proposed system. As suggested in the figure, the development aspect of the activity increases as we move away from the conceptual layer toward the manufacturing layer. Similarly, the research characteristic of the activity intensifies as we move closer to the concept layer.

In all three models of research, it is crucial for the proposed work to include unexplored problems as captured by the following remark.

Remark 1.2   Every research activity is expected to deal with an unsolved problem in its field directly or indirectly in order to meet the originality (novelty) requirement of research.

Determining if a mathematical exploration constitutes an original research activity is relatively straightforward because of the unmistakable interpretation of the mathematical steps described in Figure 1.1. All one needs to do is to verify that the proposed lemmas, theorems, and corollaries have not been proven before.

Example 1.1: Number theory

A well-known problem in number theory is concerned with the investigation of the infinitude of what are called twin primes, that is, numbers in the form of p and $p + 2$, $p > 1$, neither of which is divisible by any number except by itself or 1.* For example, 3 and 5 are twin primes as are 5 and 7, and 11 and 13. Even though the twin prime conjecture states that there are an infinite number of such pairs of primes, no proof of this conjecture has appeared to date. Clearly, one cannot prove the conjecture by finding all pairs of twin primes as this may take forever. One way to prove it would be to find a formula to represent pairs of twin primes, but constructing such a formula has eluded mathematicians so far. One formula that people have tried is $(6n - 1, 6n + 1)$ for twin primes that are greater than 3. It works for $n = 1$ since $6n - 1 = 5$ and $6n + 1 = 7$, $n = 2$ since $6n - 1 = 11$, $6n + 1 = 13$, and $n = 3$ since $6n - 1 = 17$, $6n + 1 = 19$; but it does not work for $n = 4$, since $6n - 1 = 23$ is a prime but $6n + 1 = 25$ is not even though 23 and 25 differ by 2. If the formula $(6n - 1, 6n + 1)$ had represented only twin primes even if it did not represent all of them, we would have the proof of the twin prime conjecture since there are infinitely many pairs of numbers $(6n - 1, 6n + 1)$, $n = 1, 2, 3, \ldots$.

Another long-standing, unproved statement in number theory is the Goldbach-Euler conjecture[†] that states that every even positive integer >2 is a sum of two prime numbers such as $4 = 2 + 2$, $6 = 3 + 3$, $8 = 3 + 5$, $10 = 5 + 5$, $12 = 5 + 7$, and $14 = 7 + 7$.

Although the same statement does not hold for odd integers,[‡] its proof for all even numbers has eluded mathematicians since it was introduced in 1742 in a communication between Goldbach and Euler.[§]

The investigations of these two number theory problems constitute a research activity not only because no one knows if they hold or not, but also because they are cast as precise mathematical problems. In both cases, integer addition is used to state what needs to be solved. What makes both problems difficult is that it is known that there are an infinite number of primes and we seek to find a subset of an infinite set, which itself may contain an infinite number of primes. Clearly, there are several missing pieces in these two puzzles.

The following variation of the first problem is much easier to settle: *Given two finite (possibly very large) integers n and n + 2, determine if they are both primes.*

It can take a lot of time to determine if n and $n + 2$ are primes as n becomes large. Still, it will be a stretch to construe this as a research problem since any of the well-known primality testing methods can be used to determine if n and $n + 2$ are primes. However, the problem can still be converted into a research problem by changing the focus to the speed of determining if two finite numbers that differ by 2 are prime. In this case, the research activity shifts from proving or disproving a conjecture to constructing a method or algorithm to find an answer to an instance of the conjecture that is faster than previously reported or

---

* Guy, R. K. 2004. *Unsolved problem is number theory*, 3rd ed. Springer, p. 32.
† Yang, Y. 2002. *The Goldbach conjecture*, 2nd ed. World Scientific Publishing.
‡ For example, try 27.
§ Dickson, L. E. 1919. *History of the theory of numbers.* Carnegie Institution of Washington, ch. 18, p. 421.

published methods. Indeed, this is one of the central research themes in applied mathematics and computer science, where researchers constantly try to improve the speed of solving problems by devising faster algorithms.

We can similarly verify if a problem characterized under our physical model of research has any research content by checking if the proposed experiments or observations have not been performed before and confirming that they are likely to lead to new hypotheses and laws. On the other hand, the same task becomes trickier when we use the engineering research model in Figure 1.3. This is because the steps in this model are not as precisely defined as in the previous two models. Each step may possibly spawn a number of related research activities and therefore should be carefully analyzed. One method that can be applied to determine the research content of these activities is to see if they focus on the discovery of a new technique, art, or algorithm. This criterion is routinely used in government patent offices during the examination of patent applications to distinguish between inventions that result from research activities and those generated as a result of development efforts.

The following example should help clarify how this criterion can be applied to a design project in computer science and electrical engineering.

Example 1.2

Suppose that we have a supply of electronic calculator chips each of which can be used to add two numbers up to a given precision, say 20 digits. Also suppose that each of our calculator chips has a carry-out pin that indicates if the sum of two numbers it adds takes more than 20 digits to represent. It also has a carry-in pin to receive a carry. With such chips, we can clearly design a system to add two 40-digit numbers. All we need to do is connect the carry-out pin of the first calculator chip to which we feed the rightmost (least significant) 20 digits of our two numbers to the carry-in pin of the second calculator chip to which we feed their leftmost (most significant) 20 digits as seen in Figure 1.4. In doing so, we have not relied on any new knowledge or fact. Rather, we made use of the well-known fact that the addition of two numbers may generate carries when their digits are added. We can use the same fact to build a system to add numbers with any number of digits. Therefore, there is no puzzle or a missing piece here.

Does this mean that our design effort does not involve any research? The answer is yes if we restrict our focus to the design of 40-digit adders using 20-digit adder chips. Our effort would not be considered research in this case. It is highly unlikely to attract any funding even if we increase the number of digits to 1 million as the solution readily follows from the principles of decimal arithmetic and positional number system. However, suppose that we change our focus and now seek to build a 10,000,000-digit adder inside a single chip instead. Such an adder would be very valuable to crunch prime numbers with

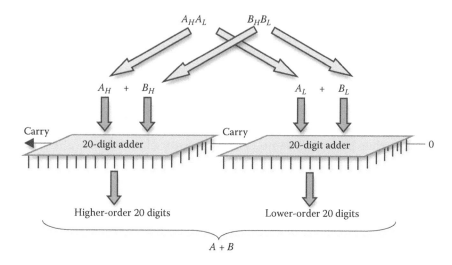

Figure 1.4 Development problem: Design of a 40-digit adder using two 20-digit adders.

millions of digits* and we have a number of formidable problems to tackle. For one, the implementation of such an adder in silicon will require over 40,000,000 1-bit adders or between 500 and 600 million transistors.† These figures are in the same range as the numbers of transistors contained in contemporary processor chips. Adding more computational features to our calculator chip will clearly make the placement of the entire system in a single chip an enormous engineering challenge. Furthermore, even if we overcome this problem, how will we feed our numbers—each of which requires 10,000,000 digits—and how will we read out the result that needs another 10,000,000 digits? Clearly, these questions would not arise if we focus only on the design of our calculator. We can easily design an electronic circuit to add two 10,000,000-digit numbers on paper using compact block diagrams. But to prototype or actually produce a chip that can perform addition on 10,000,000-digit numbers is a huge challenge that requires a fresh approach, and a new integrated circuit technology base and framework. Thus, several pieces of a puzzle are missing here, and this makes the described problem a research activity even though the design phase of the same problem does not make it so at all.

We can also modify the original design problem so that it becomes a research activity as well. For example, suppose that we want to add two numbers in an amount of time that remains constant for any number of digits we choose to use. Can this be done? If so, under what set of assumptions, can it be done? The crux

---

* The largest prime known to date has 12,978,189 digits. It was found in August 23, 2008. See http://www.mersenne.org/ (accessed December 10, 2010); WebCite, http://www.webcitation.org/5ut5UuQC8.
† It takes four 1-bit adders to build a 1-digit adder, as each decimal digit requires four bits to represent.

of these questions rests on the sequential nature of addition. We need to deal with the carries that are generated sequentially from the rightmost digit position toward the left as digits are added. Therefore, it would seem that the more digits we have, the longer it would take to complete an addition. This does not come from the difficulty of adding digits together in any given position. Rather, it comes from the temporal dependence of a sum digit on a potential carry that may have been generated as a result of the addition of the digits to its right. Without the problem of computation of carries, addition operation would be trivial, and the answer to our question would be a firm yes.

So, let us rephrase our question: Can the carries be computed in constant time? The answer to this question turns out to be yes, provided that we use some of the carry prediction methods developed in the literature. However, these methods are known to be impractical as they require operations (circuits) whose numbers of operands increase linearly with the number of digits to be added. So we need to determine if there exists a method that can compute all the carries in constant time using operations with a constant number of operands.* If not, can a new number system be devised to eliminate the need to add two numbers without generating carries? If so, would this lead to a system with a constant addition time as depicted in Figure 1.5? If not, could it be established that such a system cannot be even designed conceptually, let alone in a physical form and however we fix the number system to carry out our addition. So, again, there is a puzzle here and several pieces are missing. Therefore, the design problem we have described is an open problem and its investigation becomes a research activity.†

The foregoing discussion illustrates that research and development activities in applied sciences and engineering fields serve complementary functions. This complementary relation is often called the research and development cycle as shown in Figure 1.6, where research activities are classified into "basic research" and "applied research" categories. A basic research activity typically involves itself with why questions, whereas an applied research activity is more likely to focus on how questions. For example, "Why do objects fall?" is a basic research question, whereas "How do objects fall?" is an applied research question. Answering how objects fall will likely provide more information for designing objects that may fly than explaining why objects fall. However, they both are still relevant to building flying objects. Basic and applied research activities are

---

* This requirement is generally referred to as the constant fan-in condition in electronic circuit design.
† The complexity of addition and other arithmetic operations in the binary domain has extensively been investigated but several problems remain open. For a survey, see Pippenger, N. 1987, The complexity of computation by networks, IBM, *J. Res. Dev.* 31:2:235–243. For a more recent study that uses a speculative addition algorithm to speed up addition, see Verma, A. K. et al. 2008, Variable latency speculative addition: A new paradigm for arithmetic circuit design, in Proceedings of the Design, Automation, and Test in Europe (DATE) Conference, pp. 1250–1255.

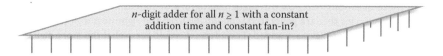

Figure 1.5 Research problem: Does a decimal *n*-digit adder with constant addition time and constant fan-in exist for all *n* > 1?

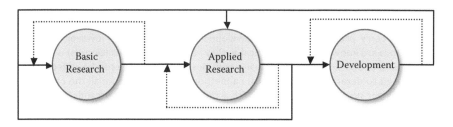

Figure 1.6 The research and development (R&D) cycle.

thus intertwined. They collectively drive development activities, which, in turn, generate new questions for basic and applied research as indicated by the feedback lines in Figure 1.6. Each activity may also feed on its own discoveries for some time. For example, research in pure mathematics often advances on its own motivating causes and interests as indicated by the dotted cycle on the left in the figure. Similarly, research activities in applied sciences may feed on their own discoveries for periods of time before they need new basic research results to move to another frontier. The same may also occur in development cycles even though the period of such self-sustaining activities may not be too long as compared to basic and applied research activity cycles.

1.3 *What makes a research activity more significant than others?*

It was earlier stated that every research activity is expected to make a tangible contribution by discovering some new facts or adding new knowledge to its field. This attribute of a research activity captures its significance or impact and leads us to the following deductive remark*:

---

* Here we postulate that the originality or uniqueness of any activity draws attention to itself.

Remark 1.3 The impact of any research activity as measured by its recognition in its own field and other fields will likely have a strong correlation with the originality of its findings.

The originality of findings published in a research article often manifests itself in the number of citations it draws from the articles in its field of study and adjoining fields. The ideas and results published in a research article may be viewed as propagating like waves in a sea of knowledge. The more original they are, the less friction they will likely experience and the more they would resonate along the way to reach more audiences. The following example should be helpful to illustrate this relation between the originality of results and their recognition and impact.

Example 1.3

Figure 1.7 plots the number of citations of one of the most widely cited research articles published in hominid research in anthropology.* At the time of this writing,[†] this article had received 300 citations on Web of Science since its publication in July 2002. It ranks third among the 2860 articles in terms of the annual rate of citations it has received and ranks seventh in terms of the total number of citations, where the highest number of citations of 446 was received by an article published in 1967.

It is instructive to examine this article in some detail to see why it has received so much attention within its field. Let us begin with its title: "A New Hominid from the Upper Miocene of Chad, Central Africa." Clearly, the word *new* in the title signals to the reader there is something in the article that has not been reported earlier in the field. The use of words such as *new* or *novel* in a title is quite appropriate as long as it is backed up by the content of the article. Of course, the title should not be misleading or give a false impression that there is something new in the article when there is not. This can disappoint the readers and may cause them to think that the author is not credible. When they encounter another article written by the same author or authors, readers will likely be skeptical and avoid reading it, let alone cite it in their own work. In research, good repute of a researcher is just about all that counts for receiving funds and getting results published.

Now let us look at the abstract of the article to further analyze the reason behind its broad recognition; key phrases are in italics.

The search for the earliest fossil evidence of the human lineage has been concentrated in East Africa. Here we report the discovery of *six hominid specimens* from Chad, central Africa, *2,500 km from the East African Rift Valley*. The fossils include a nearly complete cranium and fragmentary lower jaws. The associated fauna suggest the *fossils are between 6 and 7 million years old*. The fossils display a unique mosaic of primitive and derived characters, and constitute a new genus and species of

---

* Brunet M., Guy, F., Pilbeam, D. et al. 2002. A new hominid from the Upper Miocene of Chad, Central Africa. *Nature* 418:145–151.
[†] December 31, 2010.

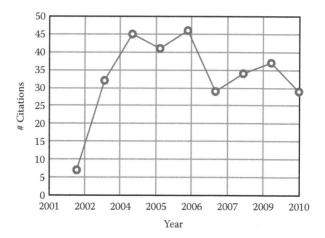

Figure 1.7 The number of citations to articles in hominid research since 2001.

hominid. The *distance from the Rift Valley,* and the great antiquity of the fossils, suggest that the *earliest* members of the hominid clade were *more widely distributed than
has been thought,* and that the *divergence between the human and chimpanzee lineages
was earlier than indicated by most molecular studies.*

Parsing the abstract, one comes away with a wealth of information that spells
out that

1. A specific number of hominid specimens were found.
2. These hominids were aged to be between 6 and 7 million years old.
3. These hominids were found at a considerably far away distance from the
   previously excavated hominid specimens to conclude a wider distribution
   of the hominid clade and an earlier divergence of human and chimpanzee
   lineages.

All these findings make a direct hit at the core of research that must be of concern and interest to other anthropologists working in the field. This is precisely
what the provided citation statistics corroborate.

Is it possible to write this abstract in such a way that the results are less
pronounced or stated without as much precision? The answer is clearly yes as the
following modified version should make this clear:

A study was conducted to search for hominids in Africa. New specimens were
found in Chad, Central Africa, and they are older than the hominid specimens that
were found earlier in East Africa. The study shows that humans and chimpanzees
had an earlier divergence than previously thought and they were more widely
distributed.

At first, this summary sounds almost identical to the actual abstract of the
article but a careful second reading reveals that there is a limited specification
of location and no information about the age and number of specimens, and this
makes a huge difference. By using numbers, the authors make the announcement

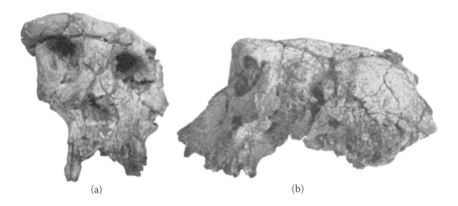

(a)   (b)

Figure 1.8 *Sahelanthropus tchadensis* gen. et sp. nov. (a) Facial view. (b) Lateral view. (From Brunet M., Guy, F., Pilbeam, D. et al., 2002, A new hominid from the Upper Miocene of Chad, Central Africa, *Nature*, 418 (6894): 145–151. With permission.)

of their contributions much more concrete and entice the reader to read on to see how these numbers are actually figured in the study.

For further clues behind the article's success, we peek into its body where the results are reported and explained. First, the authors use the introduction section brilliantly to make it clear that the conducted field study of hominids has produced concrete new findings. They emphasize that hominid specimens have been found in a wider area of African continent than previously thought. They then compare the characteristics of new hominid specimens with the hominid specimens found earlier in East Africa. Most important, they provide images of the specimens to make sure that the reader gets the idea: The findings of the article are real and based on concrete observations, not fiction (see Figure 1.8). Finally, the article is concluded with a discussion of the impact of its findings on the state-of-the-art knowledge in the field. It shows how it adds another piece to the completion of the puzzle without leaving any doubt in the reader's mind. It demonstrates that its results vary a great deal from the previously obtained results, while they are also integrated with those results to work toward the completion of the puzzle. This is what makes it as widely recognized as it has been.

How did the other articles in the field fare against this article? The statistics gathered from the Web of Science at the time of writing this text show that the average number of citations received by the 2860 hominid articles published since 1980 is 18.97 as compared to 33.33 for the hominid article reviewed. The *h*-index of these 2860 articles is 91 implying that 91 of them have received 91 or more citations. Equivalently, 96.82% of the articles received less than 91 citations. The distribution of the number of citations over the last 30 years is shown in Figure 1.9. Nearly 35% of 52,942 citations occurred in 2008, 2009, and 2010; nearly 50% occurred between 2006 and 2010 with about 1707 citations per year over the last 31 years. This indicates that the activity in the field has increased significantly between

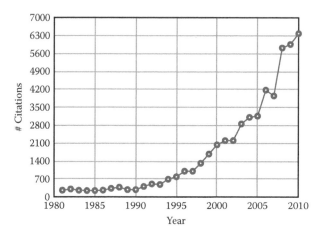

Figure 1.9 Citations to articles in hominid research since 1980.

2006 and 2010. Of the 2860 articles, 22.5% received no citations at all, that is, more than 1 out of 5 articles received not a single citation. Of course, these statistics include articles that have been published recently. If the articles published in 2009 and 2010 that received no citations through the end of 2010 are excluded, the percentage of articles receiving no citations drops to 18.74%, which is still more than 1 out of 6 articles.

It is reasonably safe to assume that the authors whose articles failed to receive any citations would rather not have written them. However, how would they know that even a single researcher in the field would not recognize their work? It is difficult to determine with certainty how many citations a given article may receive in the future. Nonetheless, by analyzing and understanding the reasons behind the success of the highly cited articles and failures of the rarely cited articles, it may be possible to write articles that are more robust for making an impact, as in the statistics given earlier. Several tools provide citation maps for researchers to view the forward and backward citations made to and by research articles. The ISI Web of Knowledge[SM] dynamically generates such citation trees up to a depth of 2. A backward and forward citation tree of depth 1 generated by this tool for the hominid article is shown in Figure 1.10. The leaf nodes on the left of the root correspond to the articles that were cited by this article. Those on the right correspond to the articles that cited this article.

To obtain a more comprehensive analysis, articles that cite a highly cited paper within a fixed window of time can be categorized with respect to their relevance to the subject of the highly cited article. The articles in a category of interest can then be compared and classified based on

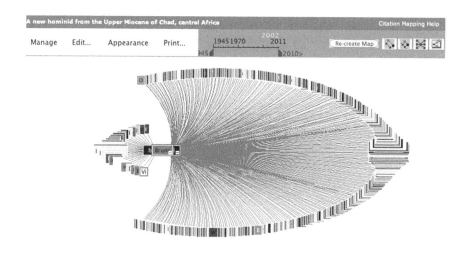

Figure 1.10 The citation tree of the 2002 article "A New Hominid from the Upper Miocene of Chad, Central Africa." (Thomson Reuters,[SM] www.science.thompson-reuters.com.)

the extent of the overlap of their results with the highly cited article. As stated in Remark 1.3, the significance of the articles in each class will likely be correlated by the degree of difference between their findings with the results of the highly cited articles. In the case of the highly cited hominid article, the 300 articles that cite it are categorized as shown in Table 1.1 with some of the papers counted in more than one field. It is seen that the largest number of citations received by the highly cited hominid article comes from multidisciplinary sciences. Furthermore, the hominid article has been cited by articles in 45 fields of research as determined by the Web of Knowledge analysis. These findings indicate that the hominid article has had a broad appeal and this accounts for the large number of citations it has received.

The hominid article has received 44 citations from articles in anthropology, which is its main field of activity. If the numbers of citations received by the anthropology articles with three or more citations (34 articles) are plotted against their relevance to the hominid article as reported by the Web of Knowledge, we obtain the graph in Figure 1.11. It is seen that the most highly cited five anthropology papers (circled numbers) that cite the hominid article are among the first one-third of articles along the relevance axis. More than two-thirds of the citing articles are more relevant to the highly cited article than these five articles. It is also worth observing that a few of the articles in the first half of the relevance space received small numbers of citations (those on the lower left-hand side in the figure).

Table 1.1 Subject Categories of the Papers Citing the Highly Cited
Paper in Hominid Research

Subject	Number of Citations
Multidisciplinary sciences	71
Evolutionary biology	55
Genetics and heredity	46
Anthropology	44
Paleontology	41
Biochemistry and molecular chemistry	25
Geosciences and multidisciplinary	21
Biology	19
Anatomy and morphology	15
Geography and physical sciences	11
Zoology	9
Cell biology	5
Ecology	5
Other	53

Source: Web of Knowledge, Thomson Reuters, www.science.thomsonreuters.com.

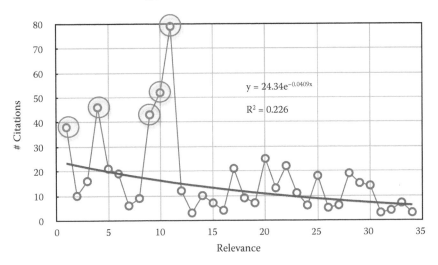

Figure 1.11 Significance of anthropology articles versus their relevance to hominid article.

These are likely to be either publications that are more recent or those that missed their marks for other reasons.

An exponential regression of the numbers of citations received by the 34 papers further shows that the number of citations received by a paper (its significance) decreases as its relevance to the highly cited paper increases. It would seem reasonable to assume that the degree of relevance of a paper to a highly cited paper provides a degree of overlap between the two and will likely determine its originality in the field. The trend line supports this assumption since the number of citations received by a paper (and hence its significance) decreases as its relevance to the highly cited paper increases. This analysis can be developed further by examining the content and subject relations between the citing papers and a highly cited paper within a pool of research papers. Such an in depth analysis can prove valuable in methodically determining the unexplored areas of the field and planning the scope and direction of a new research project if achieving high impact is among the objectives of a research project. We will revisit this question in Chapter 2.

1.4 The scientific integrity and intellectual merit of a research activity

So far we have emphasized that every research activity is expected to focus on unsolved problems. This requirement is crucial but is not sufficient to make a research activity a scientific research. For this, a research activity must be conducted using a scientific approach as captured in the following remark:

Remark 1.4 Every scientific research activity is expected (a) to formalize a set of problems within a scientific framework and (b) to use a scientific approach to solve them.

The two attributes of a scientific research activity asserted in the remark underscore its scientific integrity* and together with the originality property described in Remark 1.2, they determine its intellectual merit:

Remark 1.5 Intellectual merit = Scientific integrity + Originality.

Effectively, every scientific research activity is expected (a) to deal with scientific problems that have not been addressed or solved before, and (b) to have an original scientific approach to solve them. Most funding agencies, including the U.S. National Science Foundation (NSF) and European Research Council (ERC), require research proposals to spell out their intellectual merits.[†,‡] When neither the stated problems nor the

---

* I use the phrase "scientific integrity" to suggest that a research activity is conducted entirely within the norms of science and using scientific methods.

† National Science Foundation. 2010. NSF proposal and award policies and procedures guide. III-1.

‡ European Research Council. 2010. ERC grant schemes guide for applicants for the advanced grant 2011 call, pp. 33–34.

approach of a research proposal exhibit a tangible degree of originality and scientific integrity, it will likely be rejected by any funding agency that supports scientific research projects.

Scientific integrity also stipulates that every scientific research activity should disclose all direct and indirect relations of its results with all prior research in the field as much as possible. Avoiding such disclosures infringes upon the intellectual property of other investigators and is unethical. It also impairs the progress of science, as contributions to a field of research constitute its reservoir of knowledge from which new results are derived. The discovery of new facts and results will likely be delayed if the relationships between existing results are not completely revealed or disclosed.

To give an eminent example, Heinrich Hertz wrote in the introduction of *Electric Waves*[*]:

> While this paper was in press, I learned that its contents were not as new as I had believed them to be. The Geographical Congress of April 1887 brought Herr W. von Bezold to Karlsruhe and into my laboratory. I spoke to him about my experiments; he replied that years ago he had observed similar phenomena, and he drew my attention to his "Researches on the Electric Discharge" in vol. cxl. of Poggendorff's Annalen. This paper had entirely escaped me, inasmuch as its external appearance seemed to indicate that it related to matters quite other than electric oscillations .... In an appendix to my paper, I acknowledged Herr von Bezold's prior claim to a whole series of observations."

Obviously, Hertz did not want to leave out a citation and acknowledged it as soon as he found out that it was related his own work.

1.5   The scientific method: Common denominator of all scientific research

The scientific integrity of any research activity rests on its method of research. The following principal steps of the scientific method enforce the scientific integrity of a research activity across all scientific disciplines:

1. Observations of empirical phenomena or experimental outcomes.
2. Explanation of observations and measurements by developing hypothetical models and theories.

---

[*] Hertz, H. 1893. *Electric waves: Being researches on the propagation of electric action with finite velocity through space.* Macmillan and Co.

3. Prediction of other phenomena and events using the developed models and theories.
4. Verification of predictions by further observations and experiments.

1.5.1   Observation and experimentation phase

Scientific findings and laws originate from empirical and experimental observations. As such, every research activity is tied directly or indirectly to an observation of some empirical or experimental phenomenon. Empirical phenomena and events arise naturally on their own. On the other hand, experiments are designed and conducted deliberately to understand the behavior of objects, events, or phenomena of interest. Still, observations and measurements must be made with utmost objectivity and accuracy in both cases for the scientific integrity of the research activity in question. In particular, the design and execution of experiments must be documented as completely as possible and their outcomes must be reproducible under the documented conditions.

Empirical and experimental research activities both focus on making observations in a physical environment. The physical environment can be the entire the universe itself as in the case of experimental astronomy research, subatomic particles as in the case of experimental nuclear physics, or DNA and RNA molecules as in the case of cell biology. Thus, the goal of empirical or experimental research is to make observations or measurements in some corner of the physical universe, whether this corner is the distant galaxies or stars subject to the laws of celestial mechanics and general relativity theory, or fermions obeying Fermi–Dirac statistics in the quantum world of atoms.* Each observation or measurement requires its set of tools such as terrestrial or space telescopes to capture the images of distant stars and scanning or atomic force electron microscopes to capture the images of atoms and electrons.

Some experiments are designed to determine the physical properties of natural entities like their shapes, structures, dimensions, intensities, weights, and so on. Such experiments are commonly referred to as measurement designs. Other experiments, called process designs, aim to determine the behavior of natural or man-made objects under certain conditions and as such conditions are varied. Experimental sciences heavily rely on statistics and other branches of mathematics for designing experiments and to ensure that measurements and observations are reliable within a margin of error. Typically, observations and measurements are designed with a set of quantities or factors in mind. Some of

---

* Bellac, M. L. 2006. *Quantum physics.* Cambridge University Press.

these quantities, called dependent variables, are measured or observed as others, called independent variables, are varied. In some cases, some of the independent variables may have to be controlled to determine the cause and effect between the independent variables of interest and those to be observed. Understanding the statistical and combinatorial properties of experiments and the intricate relations between the independent and dependent variables is crucial to designing and conducting credible experiments.[*][†]

In some cases, experiments can be conducted as thought processes without making physical observations. Such experiments are called gedanken (hypothetical) or thought experiments.[‡] The history of science includes numerous examples of such thought experiments, including Galileo's thought experiments on falling objects and rolling balls, Maxwell's demon to contradict the second law of thermodynamics, Schrodinger's cat to challenge the superposition principle of quantum mechanics, and Einstein's elevator thought experiment that played a key role in the development of his general relativity theory. Typically, a thought experiment creates a scenario by which a physical behavior can be concluded without actually making a physical observation or measurement.

Designing and performing experiments is as much an art as it is a science. It involves a careful analysis of the likely outcomes and a solid understanding of the potential interferences between the experiment and the environment within which it is set up to run. All experiments carry a certain amount of error in their measurements in one form or another that must be controlled properly in order to have reliable and meaningful observations.

The following remarks by Hertz underscore the importance of designing experiments[§]:

> I have often been asked how I was first led to carry out the experiments which are described in the following pages. The general inducement was this. In the year 1879, the Berlin Academy of Science had offered a prize for a research on the following problem: Establish experimentally any relation between electromagnetic forces and the dielectric polarization of insulators.

---

[*] Montgomery, D. C. 2001. *Design and analysis of experiments*, 5th ed. John Wiley & Sons.
[†] Toutenberg, H. 2002. *Statistical analysis of designed experiments*, 2nd ed. Springer-Verlag.
[‡] Brown, J. R. 2010. *Laboratory of the mind: Thought experiments in the natural sciences*, 2nd ed. Routledge.
[§] Hertz, *Electric waves*.

And he continued:

> I reflected on the problem, and considered what
> results might be expected under favorable conditions
> by using the oscillations of Leyden jars or of open
> induction coils. The conclusion at which I arrived was
> certainly not what I had wished for; it appeared that
> any decided effect could scarcely be hoped for, but
> only an action lying just within the limits of observa-
> tion. I therefore gave up the idea of working at the
> problem; nor am I aware that it has been attacked
> by anybody else. But in spite of having abandoned
> the solution at that time, I still felt ambitious to dis-
> cover it by some other method and my interest in
> everything connected with electric oscillations had
> become keener. It was scarcely possible that I should
> overlook any new form of such oscillations in case a
> happy chance should bring such within my notice.

The "happy chance" to which Hertz was referring was the experimental
verification of Maxwell's discovery of electromagnetic waves propagating at
the speed of light, ultimately leading to the discovery of radio waves in 1887.

The significance of performing or conducting an experiment accurately should not be underestimated either. Michael Faraday, one of the
greatest experimental scientists of all times, wrote in his book *Chemical
Manipulation**:

> There are two parts in an experiment; first, it has to
> be devised; its general nature and principles are to
> be arranged in mind, and the causes to be brought
> into action with the effect to be expected, properly
> considered; and then it has to be performed ….
> There remains the mere performance of it, which
> may properly be expressed by the term *manipulation*.
> Notwithstanding this subordinate character of
> manipulation, it is yet of high importance in an
> experimental science, and particularly in chemistry.
> The person who could devise only, without know-
> ing how to perform, would not be able to extend
> his knowledge far, or make it useful …. By accurate
> and ready manipulation, therefore, an advantage is

---

* Faraday, M. 1827. *Chemical manipulation*. W. Phillips, George-Yard, Lombard Street,
  pp. iii–iv.

gained independent of that belonging to the knowledge of the principles of the science, and this is so
considerable, that of two persons having otherwise
equal talents and information, the one who manipu-
lates (performs the experiment) best will very soon
be in advance of the other.

Whether it is conducted in a physical world or as a thought process, experimental research complements theoretical studies by providing observations and data on which theoretical models thrive. Therefore, the standards for conducting experimental research ought to be as stringent as they are for theoretical research because inaccurate recording of observations can hamper the verification of scientific models and may lead to wrong conclusions.

1.5.2   Modeling phase

Experimental and empirical observations require cohesive explanations in terms of scientific models. The scientific literature includes countless such models that extend from simple linear systems to much more sophisticated nonlinear and dynamic models. Perhaps the preeminent example of scientific modeling of physical phenomena is Isaac Newton's development of classical and celestial mechanics. He pieced together one of the most enduring scientific theories of nature from his own observations and those of other scientists. Building on the findings of Copernicus, Kepler, and Galileo in his masterpiece, *Philosophiae Naturalis Principia Mathematica*, he laid out his discoveries of gravitational forces and orbital motions of planets, comets, and stars with such mathematical precision that is unsurpassed by any scientific investigation to date. He used every conceivable mathematical entity, including hypotheses, definitions, propositions, axioms, laws, lemmas, theorems, and corollaries, to develop the most comprehensive scientific theory of the forces of nature. Roger Cotes, a noted astronomer and contemporary of Isaac Newton, described Newton's work in the preface of the English version of the book as follows*:

Fair and equal judges will therefore give sentence in
favor of this most excellent method of philosophy,
which is founded on experiments and observations.
To this method it is hardly to be said or imagined,
what light, what splendor had accrued from this
admirable work of our illustrious author; whole

---

* Newton, *The mathematical principles*.

happy and sublime genius, resolving the most difficult problems, and reaching to discoveries of which
the mind of man was thought incapable before, is
deservedly admired by all those who are more than
superficially versed in these matters. The gates are
now set open; and by his means, we may freely enter
into the knowledge of the hidden secrets and won-
ders of natural things. He has so clearly laid open
and set before our eyes the most beautiful frame of
the System of the World.

As in Newton's remarkable work, theoretical investigations often seek
to build new mathematical or formal models that can be used to explain
observations in empirical and experimental studies or predict other physical or natural events or phenomena. The field of physics is particularly
filled with numerous examples of such theoretical investigations going
back as far as the mathematical equations discovered by polymaths of
ancient civilizations to explain and predict the natural events such as the
periodic behaviors of the sun, earth, and other planets. Other examples
include[*,†,‡,§]:

1. The laws of celestial mechanics and gravity discovered and developed by Galileo Galilei, Johannes Kepler, Leonhard Euler, Joseph-
   Louis Lagrange, Pierre-Simon Laplace, William Hamilton, and Carl
   Friedrich Gauss to explain planetary motions.
2. The electromagnetic and electrodynamics field theories developed
   by James Maxwell, William Thomson, and Hendrik Lorentz to
   explain and unify the discoveries of Hans Øersted, Andre-Marie
   Ampere, Michael Faraday, and Carl Friedrich Gauss on electricity
   and magnetism and the interplay between them.
3. The special relativity theory developed primarily by Albert Einstein,
   Hendrik Lorentz, Henry Poincare, Max Planck, Hermann Minkowski,
   and others to unify the electrodynamics and electromagnetic field
   theories with Newtonian laws of gravity and mechanics.
4. The general relativity theory developed by Albert Einstein, Hermann
   Minkowski, and David Hilbert to incorporate the relativistic time
   into Newton's laws of gravity celestial mechanics and to better
   predict the motions of astronomical objects.

[*] Newton, *The mathematical principles.*
[†] Bunch and Hellemans, *History of modern science.*
[‡] Bellac, *Quantum physics.*
[§] Feynman, R. 1985. *QED: The strange theory of light and matter.* Princeton University Press.

5. The quantum mechanics theory developed by Max Planck, Neil Bohr, Erwin Schrodinger, Karl Heisenberg, and others to explain the physical interactions between particles at atomic scales such as electrons and protons discovered earlier by Joseph Johnson Thomson and Ernest Rutherford.

6. The quantum field electrodynamics and gauge theories developed by Paul Dirac, Wolfgang Pauli, Richard Feynman, Franklin Yang, and others to explain the interactions between light and matter at atomic scales using a relativistic approach.

As in each of these historically important discoveries, developing theoretical models to explain empirical or experimental observations and measurements plays an essential role in the dynamics of scientific progress. Each new model replaces its predecessor to eliminate an inconsistency, anomaly, or inaccuracy in explaining some physical observation or measurement. It is then used to predict new physical phenomena until it fails and is replaced by another model. In the meantime, some models stand the test of time and become scientific laws and theories as they are tested and verified over long periods of time. Still, all scientific laws have a certain nonvanishing probability that they may be broken one day by some new natural phenomena or observation. Indeed, this is what constantly shifts scientific frontiers in all fields of research and generates new problems for scientists to tackle.

1.5.3   Prediction phase

In this third phase of research, theoretical models developed during the modeling phase are often converted into computationally more effective numerical, statistical, and simulation models to predict what other and yet unobserved events and phenomena may occur. Thus, the prediction phase allows researchers to test out established theories and laws against new experimental and empirical findings. The choice of prediction model depends on the representation used to record the observations and measurements as well as efficiency considerations. Often the original representation of observations and measurements is transformed to another space of computations to reduce the amount of processing to predict the derived outcomes or results. This transformation makes the theoretical models developed during the modeling phase more applicable and computationally more effective. Numerous mathematical techniques such as continuous and discrete transforms, and probabilistic and statistical prediction techniques are used to simplify the analysis and design of physical systems by mapping problems between different domains of interest.

1.5.4 Verification phase

The last phase of a scientific research activity is concerned with the verification or validation of the predictions foreseen by the models developed in the modeling and prediction phases. The verifications are carried out by way of further observations, experiments, simulations, and test beds. Any disagreements between the predicted outcomes and empirical or experimental observations would require repeating the observations, experiments, and simulations as well as questioning the validity of the models and predictions.

These four phases of a scientific study are repeated as many times as needed to strengthen the correlations between predictions and actual results as depicted in Figure 1.12. Not all four steps are necessarily invoked in every research activity but the continuity and integrity of scientific investigation requires that these principal methods of inquiry and research be employed within each field of research activity. Furthermore, these four steps may take on different forms in different fields of study and be refined further to meet the demands of the scientific research in question. It should also be emphasized that individual research projects may focus on only theoretical aspects of an open field of problems or only its experimental aspects. However, the problems in the field cannot be solved completely until the discovered theoretical and experimental results are fused together through the modeling, prediction, and verification steps of the scientific method.

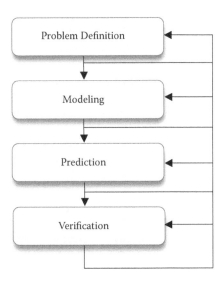

Figure 1.12 The four steps of the scientific method.

1.6 Empirical, experimental, and theoretical research

In Section 1.2, we have introduced mathematical, physical, and engineering models to describe the style and structure of research used in various scientific disciplines. Among these three models, the physical model most closely resembles the style of scientific method of research that has been described in the previous section. Does this mean that the mathematical and engineering models of research deviate from the scientific method? The answer is clearly no, and in fact, all scientific disciplines, including mathematics,* make use of varying degrees of empirical, experimental, and theoretical methods of scientific research. Such methods are employed within more refined layers of research in one form or another, and may not be as exposed or explicit as in the physical model of research. For example, Euclidean geometry, a well-established branch of mathematics, relies on postulates that are derived from drawings of lines and circles on a plane, which are clearly empirical in nature. Likewise, experimental and empirical investigations and scientific laws are interwoven into the various layers of the engineering research model that has been described in Section 1.2.

The bar graph in Figure 1.13 depicts a likely relation between various disciplines of research and their use of such methods as well as the extent of mathematical content and symbolism they employ. The graph was obtained by classifying research articles published between 1980 and 2010 into various research fields using Scopus®, an online database. The articles were classified into 13 research fields and 4 research categories using the keywords *empirical*, *experimental*, *theoretical*, and other descriptive words to characterize particular aspects of these fields. The research fields are displayed from top to bottom in descending order of the degree of their use of mathematics as indicated by the percentage figures on the right-hand side of the figure. Each scientific field was further analyzed in terms of how their research methods are distributed among the three categories we used. The percentages of distribution of each field into these categories are marked inside the bar segments of its row in the figure. The light gray segment in each horizontal bar represents the percentage of research articles that did not show up in our search in any of the empirical, experimental, or theoretical categories. These regions can be viewed as belonging to two or more of the three categories and thus labeled as

---

* There is an emerging field of experimental mathematics that focuses on numerically testing mathematical assertions and searching for patterns in mathematical structures by paper-and-pencil methods and computer programs. See A. Baker, 2008, Experimental mathematics, *Erkenntnis*. 68:331–344. Also see http://www.expmath.org/expmath/philosophy.html. (accessed December 5, 2010); WebCite, http://www.webcitation.org/5ul4YXrpl.

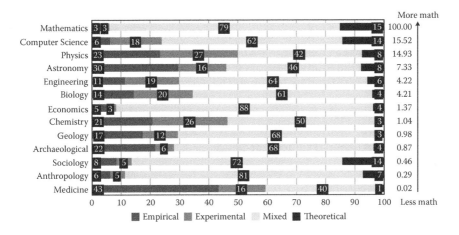

Figure 1.13 Empirical relation between fields of research, approach, and mathematics.

mixed in the figure, but how the three categories were mixed was not determined in our classification.

The following conclusions can be drawn from our study:

In so far as the use of mathematics in a scientific article, research articles in mathematics expectedly top the list with those in computer science and physics ranking second and third, and those in medical sciences occupying the last position.

A positive correlation exists between the use of mathematics and theoretical content of the research articles in physical sciences.

The empirical, experimental, and theoretical aspects of the research articles in the mixed regions will likely vary with the field. For example, in mathematics research, the theoretical aspect of most of the articles would likely dominate their empirical and experimental aspects with a few exceptions.* Along the other extreme, clinical (empirical), and experimental aspects of medical research articles within the mixed region are expected to dominate their theoretical aspect as confirmed by our empirical study.

Some fields such as biology and chemistry use the symbolism of mathematics to formalize their observations and findings in terms of formulas and graphs. In other fields such as physics, computer science, and engineering, the axiomatic and analytic machinery of mathematics is used to build theories, verify the behavior of systems, and prove the correctness of operations of software and hardware systems. There have been efforts to axiomatize physical sciences after David Hilbert provided an axiomatic representation of Euclidian geometry and asked if physics

---

* Baker, "Experimental mathematics."

can also be axiomatized at the turn of the 20th century. Despite significant efforts by John von Neumann, Garret Birkhoff, and others,*†‡ Hilbert's question, known as his 6th problem, remains open.

Our empirical study also suggests that research articles in social sciences expectedly tend to use much less mathematical symbolism and abstraction than those in mathematical and physical sciences. However, this does not make them any less theoretical. In fact, our study shows that the percentage of theoretical research articles in sociology matches the percentage of those in computer science.

1.7   Interplay between basic and applied research

It can be said that the scientists of the past had a distinct advantage to keep up with scientific developments as science was not as advanced and branched out as it is today. It was not even separated into basic and applied research fields. Still, the work of the polymaths of the past demonstrates that the impact of their contributions was greatly enhanced by their ability to conduct research in a number of fields.

Today, basic research refers to a scientific investigation in a field where the goal is to answer an open question without an explicit desire to develop tools, methods, and products to solve a practical problem. Such open questions can arise in nearly all scientific fields as illustrated by the following examples:

- How big is the universe? Does it end somewhere?
- How many galaxies are there in the universe?
- Why do living organisms die?
- Why do some plants live longer than others?
- When did life begin?
- Is there infinitude of twin primes?
- Is light a wave or a particle?

Most of these questions, if not all, often arise in a researcher's mind without any motivation to solve a practical problem or address a need. This is not to say that answering such questions will never lead to anything useful. Rather, they are often not posed with an intention of addressing any such need or use.

---

* Neumann, J. V. 1955. *Mathematical foundations of quantum mechanics*. Princeton University Press.
† Birkhoff, G., and Neumann, J. V. 1936. The logic of quantum mechanics. *Annals of Math.* 37:823–843.
‡ Redei, M., and Stoltzner, M. 2006. Soft axiomatization: John von Neumann on method and von Neumann's method in the physical sciences. In *Intuition and axiomatic method*, ed. R. Heuber and E. Carson, pp. 235–249. Kluwer.

In contrast, the following are examples of practical or applied research questions:

- How can humans colonize other planets?
- How can flu epidemics be controlled?
- Can human genetics be altered to fight diseases?
- How can congestions in thoroughfares be reduced?
- How can numbers be divided faster?
- How can a car be designed to use less fuel?
- Can materials be designed to hear and see?

Here, the meaning of a practical problem may be a little vague or blurred, but perhaps inevitably so. Broadly speaking, the word *practical* implies that the solution of the problem in question can relieve a need or help alleviate performing a task. Most tools and products perform such a function and have been perfected as a result of many thousands of hours of applied research, development, and manufacturing efforts. It is difficult to imagine that basic research alone would have led to count-less many tools and products without which we can hardly function today. However, this does not mean that basic research is irrelevant to solving practical problems or alleviating needs. Quite the opposite, as mentioned earlier; theoretical studies in electrical science and engineer-ing during the past two centuries played a pivotal role in the develop-ment of electronics technologies of today. A more contemporary example is the relationship between number theory and cryptography. Number theory is one of the oldest branches of mathematics dating back to ancient civilizations of Mesopotamia, Egypt, and Greece. Many great minds of mathematics from Pythagoras, Euclid, and Eratosthenes of ancient Greece to Euler, Fermat, and Gauss have raised and answered such basic ques-tions as divisibility of numbers, finding the greatest common divisors of numbers, factorization of numbers into powers of prime numbers, den-sity of primes, among others. These and many other problems in number theory dealing with intrinsic properties of numbers have been raised and studied mostly in a theoretical context without being explicitly tied to the resolution of some practical problem or need for centuries. However, with the arrival of the information age, many intricate lemmas and theorems developed in number theory over two millennia have suddenly found a great utility in reliable and secure transmissions of digital text, audio, and video files through remarkable contributions of mathematicians and com-puter scientists who work in encryption research today.

It should be noted that basic research is not limited to mathematical investigations. It can be conducted in physical and social science as well but the classification of the motivation for doing so, that is, whether or not it is done to help solve a practical problem, becomes more difficult. For

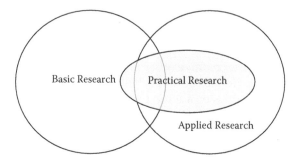

Figure 1.14 Possible relation between basic and applied research.

example, in material science, basic research projects may employ solid-state physics to discover new materials with better mechanical, electrical, magnetic, thermal, or optical properties. If discovered, such new materials may potentially be useful in making better products and building better physical systems, where better can mean more durable, stronger, lighter, smaller, faster, and so forth. Therefore, in a way, basic research in material science is a little closer to applied research than basic research in most fields of mathematics. It is even possible to view such a research activity as applied research given that it deals with the problems of building better materials by some measure or another that are bound to find applications and help solve practical problems along the way.

Figure 1.14 displays a possible classification of research projects into basic and applied research categories. The intersection of the two categories of research suggests the possibility of research projects that carry the signatures of both basic and applied research as in the materials science research example. The figure also depicts a subregion of "practical research" that is unevenly distributed between basic and applied research. Practical research deals with product-oriented issues and problems and is closer to the engineering side of research than the basic sciences side. Edison's invention of the incandescent light bulb is an example of practical research. He was set on building a product that would provide light through electricity and help people see after sunset with all the basic and applied research results available to him in 1879. On the other hand, Newton's investigation of the laws of gravity and celestial mechanics, Faraday's experiments on electricity and magnetism, Maxwell's seminal work on electromagnetic waves, and Hertz's investigations to demonstrate that electromagnetic waves actually exist are all landmark examples of basic research that led to life-changing and practical outcomes over time. A well-known example of applied research that had a similar impact and led to the proliferation of the information age and Internet technologies is the World Wide Web (WWW) protocol that was proposed by Tim Bernes-Lee and others at CERN to simplify the distribution of ideas, procedures, and results within the physics research community across the globe using hypertext documents.* The exponentially growing universe of the Internet has certainly transcended the initial practical goals of the project by several orders of magnitude, demonstrating that applied research just as basic research can indeed be beneficial beyond its intended practical goals.

More realistically, funding agencies view basic research as scientific investigations that may return their investments in a very long period, and applied research as projects that can help solve practical problems within a few years. Those research projects that fall in between the two categories are generally expected to lead to practical outcomes within a decade or two. The perspective of funding agencies on basic versus applied research will be further discussed in subsequent chapters.

1.8   Role of technology in scientific research

During a talk at the California Institute of Technology more than 50 years ago, Richard Feynman, a Nobel Prize winner in physics and one of the visionary scientists of the 20th century, suggested that designing systems in an atomic scale will not violate any laws of physics. He proposed to design and build a variety of biological and chemical nanoscale devices and systems using atoms as their primitive building blocks. He stated[†]:

> The principles of physics, as far as I can see, do not speak against the possibility of maneuvering things atom by atom. It is not an attempt to violate any laws; it is something, in principle, that can be done.

In 1965, six years after Feynman's talk, Gordon Moore, another visionary, a chemist by training and engineer at work, made a prediction that the number of transistors in computer chips would double every year for a foreseeable future.[‡] In his 1965 article, he wrote:

> The complexity for minimum cost components (transistors) has increased at a rate of a factor of two per year. Certainly over the short term this rate can be expected to continue, if not to increase. Over

---

[*] Berners-Lee, T. 2010, November 22. Long live the Web. *Scientific American*, http://www.scientificamerican.com/article.cfm?id=long-live-the-web.

[†] Feynman, R. 1960. There is plenty room at the bottom. *Engineering and Science Magazine* 23(5).

[‡] Moore, G. 1965, April 19. Cramming more components onto integrated circuits. *Electronics Magazine* 38:114–117.

the longer term, the rate of increase is a bit more
uncertain, although there is no reason to believe it
will not remain nearly constant for at least ten years.

This prediction has turned out to be too optimistic and in 1975 Moore
revised it to doubling the number of transistors every two years since 1959,
the year Feynman gave his talk on manipulating atoms.* Feynman and
Moore clearly shared the same vision to build systems at atomic scales.
As a physicist, Feynman knew that the laws of physics known at the time
would not stop him and others from manipulating atoms. He asserted that
if physicists could build electron microscopes 100 times more accurate,
biologists and chemists can solve their research problems by watching
DNA, RNA, and carbon molecules, and this may then lead to technologies to manipulate atoms biologically or chemically. Moore had a more
pragmatic view in that he knew from the data he gathered, it would take
some time to reach the atoms in germanium and silicon substrates, but if
the technology he was helping to develop had kept up as he predicted,
this would eventually happen.

Much happened since Feynman and Moore made their predictions.
Moore's prediction is now known as Moore's law. The 45-nanometer
barrier has been crossed with the feature sizes in today's microprocessor chips. Intel announced that it will market new microprocessor chips
using a 32-nanometer wafer technology by early 2011 and further reduc-
tions in feature sizes obeying Moore's law are anticipated. The graph in
Figure 1.15 depicts the transistor counts, feature sizes, and clock rates of
Intel processors in binary log scale since the announcement of the Intel
8080 processor in 1974.[†] The trend lines for the three curves suggest that
the transistor counts and clock rates have been increasing exponentially
while the minimum feature sizes have been decreasing exponentially. The
slope of the transistor count trend line indicates that the number of transistors in Intel processor chips has doubled every $\log_2 2/0.4955 = 1/0.4955$
$= 2.01$ years, which is pretty to close to what Moore predicted in his 1975
article.[‡] This figure represents the average number of years during which
the transistor count has doubled over increasing chip sizes. Using the
minimum feature size, which specifies the width of the smallest feature
that can be created on a chip, will likely give a better idea about how
Intel's processor lithography and wafer technology has done over time.
The inspection of the slope of the feature size trend line shows that the

---

* Moore, G. 1975. Progress in digital integrated electronics. *Technical Digest*, IEEE
  International Electron Devices Meeting 21:11–13.
[†] The data for the figure are taken from http://download.intel.com/pressroom/kits/
  IntelProcessorHistory.pdf.
[‡] Moore, "Progress in digital integrated electronics."

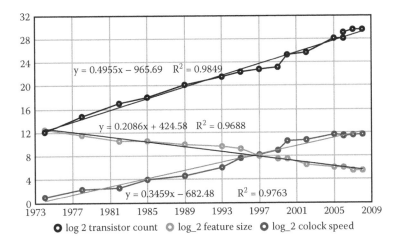

Figure 1.15 Transistor count, feature size, clock speed for Intel processors (1974–2010). (Transistor count, feature size, and clock speed values are compiled from http://www.intel.com/technology/timeline.pdf and the Web site http://www.intel.com/pressroom/kits/quickrefyr.htm.)

minimum feature size has been cut in half every $\log_2 2/0.208 = 1/0.2086$ = 4.79 years. This makes sense if we assume that the area of a transistor grows quadratically with the feature size. This assumption follows from the fact that, if we reduce the feature size from say f to $f/2$, we can then theoretically fit four transistors into an area of one transistor. By the slope of the transistor count trend line, quadrupling the number of transistors would take $\log_2 4/0.4955 = 2/0.4955 = 4.04$ years as compared to 4.79 years, which we have obtained from the slope of the feature size trend line. The discrepancy between the two figures would suggest that either the slopes in our trend line regressions are not very accurate or that there has been more innovation in transistor layout technologies than just reducing the minimum feature size to fit more than four transistors into an area of one transistor when the feature size is reduced in half. Given that the R^2 values are pretty close to 1 in all three regressions, it is more likely that the latter statement applies. Using a similar analysis indicates that the clock rates have doubled every 2.89 years.

To be sure, the transistor represents an incredibly robust concept and a physical device that has proved itself as a resilient vehicle to design systems down to the nanometer scale in 60 years of its life. In a way, a deliberate engineering research approach conducted relentlessly at Intel and other microprocessor and memory chip manufacturers to build wafers with more and smaller transistors over a stretch of time has succeeded to penetrate into the atomic world described by Feynman. Yet, Feynman's vision of "plenty of room at the bottom" goes far beyond laying out transistor circuits at an atomic scale to create microprocessor and memory chips. It proposes to replicate nearly every operation that can be performed and every system that can be built within our own visual space—a billion times larger than the space where atoms interact. This not only includes electrical systems but mechanical, biological, and chemical systems as well. For many scientists and researchers, this may sound like borderline science fiction, and in the end it may indeed turn out to be so, but government funding agencies around the globe are pouring billions of dollars into nanotechnology research. Some of Feynman's predictions have begun to happen in nanotechnology research institutes and laboratories. His wish to have more accurate and higher resolution electron microscopes did eventually come true with the invention of the scanning tunneling microscope (STM) by Gerd Binnig and Heinrich Rohrer in 1981[*] and the atomic force microscope (AFM) by Gerd Binnig, Calvin Quate, and Christoph Berger in 1986.[†] In their Nobel lecture, Binnig and Rohrer stated:

> Perhaps we were fortunate in having common training in superconductivity, a field which radiates beauty and elegance. For scanning tunneling microscopy, we brought along some experience in tunneling and angstroms, but none in microscopy or surface science. This probably gave us the courage and light-heartedness to start something, which should "not have worked in principle as we were so often told."

The use of these nanometer range microscopes has since led to significant new discoveries in nanotechnology.[‡] As predicted by Feynman, researchers in chemistry, material science, and biology have begun to investigate the structure and forms of organic and inorganic nanomaterials and discovered a range of nanostructures like nanotubes, nanorods, nanowires, nanorings, nanocages, quantum dots, quantum wells, and other carbon-based molecular structures such as fullerines, buckyballs, nanocrystals, and nanographenes.[§]

Current nanotechnology research seeks to understand how such nanostructures form and explore various biological and chemical self-assembly processes to devise techniques for building and manufacturing

---

[*] Binnig, G., and Rohrer, H. 1986. "Scanning tunneling from microscopy: From birth to adolescence." Nobel lecture.

[†] Binnig, G., Quate, C. F., and Berger, C. 1986. Atomic force microscope. *Phys. Rev. Lett.* 56:930–933.

[‡] Erokhin, V., Ram M. K., and Yavuz, O. *The new frontiers of organic and composite nanotechnology.* Elsevier.

[§] Williams, L., and Adams, W. 2007. *Nanotechnology demystified.* McGraw Hill.

systems on a nanoscale as Feynman had envisioned. As such techniques are discovered, they will have to be subject to measuring up to such rigid engineering performance metrics as size, speed, capacity, reliability, and durability, as have all other engineering systems designed and built by humans up to the present.

If nanotechnology research succeeds, it will have far-reaching consequences in nearly all aspects of our lives from health to travel much the same way as any ubiquitous technology that has been developed before. It will also likely impact the scientific research in physical sciences with new experimental sciences emerging in nanoscales and possibly new theories to model and predict the behavior of nanosystems and devices. Realizing this huge potential of nanotechnology, the government and private research funding agencies around the globe have launched nanotechnology research programs to promote and guide nanotechnology research. Japan, France, Germany, and Great Britain began funding nanotechnology early on, dating back to the 1980s. The European Union Framework Programs have been funding nanoscience and technology research since the mid-1990s. The U.S. government has been funding nanoscience and engineering research on a national scale since it launched the federal government's National Nanotechnology Initiative (NNI) in 2001. This initiative cuts across several U.S. departments and agencies, including the National Science Foundation (NSF); Departments of Defense, Energy, and Transportation; National Institutes of Health; and the National Aeronautics and Space Administration.[*] As narrated next, a recent NSF report[†] suggests that these funding programs have helped nanotechnology research move through two generations of nanotechnology products during the first decade of 21st century with two more generations to come:

> **First generation (2001+):** Passive nanostructures that include dispersed and contact nanostructures such as aerosols and colloids, and products incorporating nanostructures that include nanostructured coatings, metals, polymers, and ceramics.

> **Second generation (2005+):** Active nanostructures such as bio-active drugs and other bio-devices, new chemical and electronic nano devices incorporated into microscale systems.

[*] United States National Science and Technology Council Committee on Technology, Subcommittee on Nanoscale Science, Engineering and Technology. July 2000. National Nanotechnology Initiative: The initiative and its implementation.

[†] Roco, M. C. 2007. National Nanotechnology Initiative: Past, present, future. In *Handbook on nanoscience, engineering, and technology*, 2nd ed. Taylor & Francis.

Third generation (2010+): Systems of 3-D nanostructures such as multiscale self-assembling biological and chemical systems, and nanoscale electromechanical systems.

Fourth generation (2015+): Multifunction and heterogeneous molecular nanosystems.

At the time of this writing, the current state of the art in nanotechnology seemed to be on track with this classification as several studies using 3-D nanostructures being reported in research publications.*

1.9   Summary

In this chapter, we surveyed the progress of science and technology over its history of 4000 years, described the ingredients and methods of scientific research, and discussed the interplay between basic and applied research. We have demonstrated that the integrity of science hinges on the intellectual merit of its methods. Its successful progress depends on developing theoretical and empirical models to explain a natural or physical behavior, or a mathematical or logical abstraction, and predict other natural, physical, or abstract behavior. Research proposals that meet or exceed these fundamental characteristics of scientific research are almost certain to succeed and those that do not are likely to fail. In the next chapter, we shall further elaborate on these fundamental characteristics of scientific research and demonstrate how they impact the content of research proposals and timing of their submissions for funding.

1.10   Bibliographical notes

We have benefited from the various references cited in the footnotes in our account of the history of scientific discoveries. The reader interested in the history of science and chronology of significant scientific developments may refer to the following texts:

Applebaum, W. 2005. *The scientific revolution and the foundations of modern science.* Greenwood Press.
Bunch, B., and Hellemans, A. 2004. *History of modern science and technology.* Houghton Mifflin Co.
Sedgwick, W. T., and Tyler H. W. 1917. *A short history of science.* Macmillan.

---

* An ISI Web of Knowledge search of "3D nano" for years returned 292 research articles. Accessed November 21, 2010.

1.11 Questions

1.1 List the five most significant scientific discoveries in history with respect to your own criteria and explain if there is any relationship between them.

1.2 Is it possible to alter the chronology of scientific progress? In other words, could science have progressed differently than the way it has?

1.3 What happens if a theoretical model fails to explain a physical measurement? Should it be immediately replaced by another model?

1.4 Can science progress by making observations and predictions only? Why is modeling important in scientific research?

1.5 Are a large number of citations that a research article receives always an indicator of original research? Are a small number of citations an indicator of unoriginal research? What other factors determine the number of citations received by a research article?

1.6 What are the highly cited research articles in your field of research? What makes them widely cited?

1.7 If you conduct a study similar to the hominid paper citation study given in the text, would you also predict that there will be a negative correlation between the relevance of the papers to a highly cited paper in your field and their citation counts?

1.8 Why is experimental research important? Can physical experiments be replaced by thought experiments in all scientific investigations?

1.9 Would you rather work within the same area of research problems or switch between different areas of research?

1.10 Classify the research articles in your field of research into empirical, experimental, and theoretical groups. How are your articles distributed into the three groups?

1.11 How are applied research and experimental research related? Is every experimental research necessarily an applied research? Should every applied research include an experimental component?

1.12 Give an example of a basic research project that may be practical, and an example of an applied research project that may be impractical.

1.13 What role do you think mathematics plays in scientific research? Does using mathematics in a research project make it more scientific? Should every research project have a mathematical model to succeed?

1.14 What are the significant technologies that have been credited to the research results in your field of research? Do you see other such technologies emerging from your field of research during the next decade or so? Do you have specific ideas to write a research proposal with such technological applications in mind?

1.15 If your field of research is theoretical and you think that the funding levels in your field are declining, how will you react? Will you consider switching your field of research or consider collaborating with researchers in more applied fields of research?

chapter two

Factors impacting outcomes of proposals

There is no doubt that all researchers feel passionate about their research and would like to be competitive in getting their ideas across. They would like to be able to get their articles published in highly ranked archival journals and transactions. They would yearn to write proposals that can stand against the criticisms of the toughest researchers in their fields of research. Yet the odds of winning an award, in the best of circumstances, are a mere fraction of all the proposals submitted to a funding agency. Proposals are rejected for a variety of reasons that include lack of funding, mismatch between the focus of the proposal and funding priorities, lack of potential for broader impact, little or no intellectual merit, infeasibility of the ideas, and limited potential for practical applications. These reasons are methodically used in the evaluations of proposals by nearly all agencies that fund scientific research projects around the globe. A less than perfect showing in any of these criteria will often be more than enough for a proposal to get declined. Still some researchers beat the odds and get their proposals funded more consistently than others. Clearly, there should be some art and possibly even some science behind writing proposals that makes such prolific researchers to succeed.

In this chapter, the aim is to (a) further describe and analyze the two most significant criteria: intellectual merit and broader impact by which research proposals are judged at most funding agencies; (b) provide concrete suggestions to make proposals more robust against potential vulnerabilities that may be exposed by a systematic application of such criteria; (c) offer simple tests that can be used to decide whether a proposal is ready for submission; and (d) illustrate how bibliometric databases and tools may be used to track critical trends and changes and fast breaking papers in a research field to make sure that the ideas put forward in a research proposal remain at the forefront among likely competitors in the field.

2.1 Intellectual merit: An essential ingredient for all research proposals

In Chapter 1, we characterized the intellectual merit of a research proposal by the following three attributes:

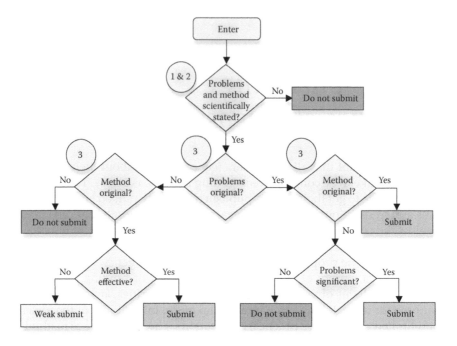

Figure 2.1 Decision diagram for intellectual merit test.x

1. A set of problems formalized within a scientific framework
2. A scientific approach designed to solve the formalized problems
3. A scientific demonstration that the problems, approach or both are new

The first two items underscore that a research proposal should rely on scientific principles to tackle a set of problems in order to have intellectual merit. The third item dictates that such a research proposal should introduce new ideas either through the problems it seeks to work on or through the scientific techniques it introduces. These requirements are the bare minimum that a research proposal must meet in order to succeed. The originality of the proposed approach and problems serves as evidence for the potential success and scientific impact of the proposed research. Figure 2.1 describes a decision process that may be used to determine if a research proposal is ready for submission for funding. Unless the first two conditions hold, the submission is not recommended. When they do, and if both the proposed method and problems are original, the submission is recommended. If either the proposed problems are new or the proposed method is new but not both, the submission option must be carefully weighed. If the problems are new but the method is not,* the significance of the problems in

_____

* Using known techniques to solve new problems is common in scientific research.

advancing the field must be clearly established. Similarly, if the problems are not new then the effectiveness of the new method[*] for their potential solutions must be clearly established. In this case, it is assumed that the problems are known to be significant in the field. In the previous case, it is assumed that the method was previously shown to be effective for solving similar problems. Otherwise, the flowchart can be extended to test these two cases as well. Finally, if neither the proposed problems nor the method are original, submission is not recommended.

2.1.1 The U.S. National Science Foundation's intellectual merit criteria

The U.S. National Science Foundation (NSF) provides explicit intellectual merit criteria for evaluating proposals. In the most recent *Grant Proposal Guide*, NSF specifies its intellectual merit criteria as follows[†]:

> How important is the proposed activity to advancing knowledge and understanding within its own field or across different fields? How well qualified is the proposer (individual or team) to conduct the project? (If appropriate, the reviewer will comment on the quality of prior work.) To what extent does the proposed activity suggest and explore creative, original, or potentially transformative concepts? How well-conceived and organized is the proposed activity? Is there sufficient access to resources?

The first question in NSF's intellectual merit definition emphasizes the originality and significance of the proposed research as it has been done here as well. The definition also makes an implicit reference to the scientific integrity of the proposed research with the question: How well-conceived and organized is the proposed activity? It additionally raises questions about the qualifications of the proposers and their ability to have access to the resources to conduct the proposed research. Furthermore, it questions if the proposed research explores how its main ideas can possibly transform the conventional scientific methods and theories into new frontiers of research and discoveries. These new additions to the intellectual merit statement seem to be designed to merge the intellectual merit and "broader scientific impact" together.

Answering the questions in NSF's intellectual merit requirement can be tricky. For example, how should the following question be answered?

---

[*] Using new techniques to solve open problems is also common in scientific research.
[†] NSF merit review criteria. 2010. In Grant Proposal Guide, chap. 3.

How important is the proposed activity to advancing knowledge and understanding within its own field or across different fields? Should we just state that the proposed activity is new; therefore it should advance the knowledge in the field, and consequently it is important? Though it might seem logical to do so, such a circular response misses the point of the question. The question already assumes that the proposal ought to have new ideas and goes one step further. It suggests that the proposal should also provide some insight or evidence by which the significant problems that remain open in the proposal's field or some related field can be solved or some significant progress can be made in that direction. Effectively, this question probes the scientific impact* of the proposed research and expects to see a tangible connection between the proposed activity and the solutions to the key problems in the field. Therefore, only stating that the proposed activity has not been undertaken before without tying it with the key problems in its field would make the proposal fail with near certainty.

The next question is more closely related with the intellectual capacity of the proposer for carrying out the proposed research: How well-qualified is the proposer (individual or team) to conduct the project? (If appropriate, the reviewer will comment on the quality of prior work.) In a way, the question is trying to gauge the intellectual merit of the proposed research by the scientific qualifications of its proposer. It seems to assume that one of the following two statements holds:

- If a proposer has carried out credible research projects in the past, then the proposed work will likely have intellectual merit.
- If a proposer does not have a good research track record, then the proposed work will not likely have intellectual merit.

It is possible for both statements to be false. A good research track record of a proposer is not a guarantee that it will remain so forever. Likewise, not having a good research track record is not a precursor for a proposal without intellectual merit. However, the NSF's intellectual merit criteria include this question to increase the probability of making the right decision. It is logical to assume that proposers with better research track records are statistically more likely to succeed. Still, the question does not completely rule out the possibility that a proposal can have intellectual merit in and of itself, as the statement in parentheses would suggest. Furthermore, in certain cases, there may be very limited

---

* Here, by the scientific impact of a research proposal, we refer to the potential recognition of the results that the proposal will likely produce if it is carried out. This should not be confused with its broader impact, which is tied with the proliferation of the proposal's ideas in new directions, and other potential applications and benefits that may be generated if it is carried out.

information about the track record of a proposer's prior research work. This is particularly true for a junior researcher or even a senior researcher who is switching to a new field. In such cases, the following question in NSF's intellectual merit statement would be more relevant: To what extent does the proposed activity suggest and explore creative, original, or potentially transformative concepts? This question provides proposers an opportunity to demonstrate the creativity, originality, and potentially transformative qualities of their ideas within the current proposal, and escape the scrutiny of their track records as much as possible. Nonetheless, a reviewer always has an opportunity to bring down a proposal by finding problems either with the track record of the proposer or the lack of creative ideas and transformative concepts in the proposal or both. In all likelihood, most reviewers would not allow a proposal to pass unless both conditions are met. This is despite the fact that the NSF intellectual merit criteria hint at the possibility of relying only on the proposal under the circumstances mentioned earlier. Therefore, to increase its probability of success, a proposal must meet both criteria as much as possible.

In particular, the phrase "transformative concepts" in the question is designed to reinforce the scientific broader impact of the proposed research and emphasize that it should be described as specifically as possible. It suggests including a plan to explore and possibly pursue the potential implications of the expected results in the proposed activity itself rather than just showing that there is potential for the results to impact other fields of study. Thus, NSF's intellectual merit criteria require a more substantive discussion of the potential scientific impact of the proposed research in other fields of scientific and technological activity, thereby ultimately probing its broader impacts.

Finally, the NSF's intellectual merit criteria include a question on the feasibility of the proposed activity: How well-conceived and organized is the proposed activity? Is there sufficient access to resources? The intent of this question is to make sure that the proposal includes a well-developed plan that factors in all the requisite resources needed to successfully carry out the proposed activity. Indeed, asking for the right resources and matching them with no-cost or in-kind resources can make a difference between winning and losing an award for certain proposals. We will discuss how resources should be determined and used effectively during the conduct of research projects in Chapter 4.

To help decipher the NSF's intellectual merit criteria further, we have coded them into a decision table as shown in Figure 2.2 with the following correspondence:

Significance of Proposed Activity—How important is the proposed activity to advancing knowledge and understanding within its own field or across different fields?

Originality of Proposed Activity—To what extent does the proposed activity suggest and explore creative, original, or potentially transformative concepts?

Feasibility of Proposed Activity—How well-conceived and organized is the proposed activity? Is there sufficient access to resources?

Qualification of Proposers—How well qualified is the proposer (individual or team) to conduct the project? (If appropriate, the reviewer will comment on the quality of prior work.)

A careful examination of the flowchart reveals that the first two questions in the intellectual merit test shown in Figure 2.1 have been compressed into a single question at the top of the flowchart in NSF's intellectual merit test. In other words, the proposed activity significant question in Figure 2.2 subsumes the method and problems scientifically stated and problems original questions in Figure 2.1. The effectiveness of the method and significance of the problems in the intellectual merit test in Figure 2.1 have been replaced by feasibility and proposer's qualifications in NSF's intellectual merit test. Of these two criteria, the

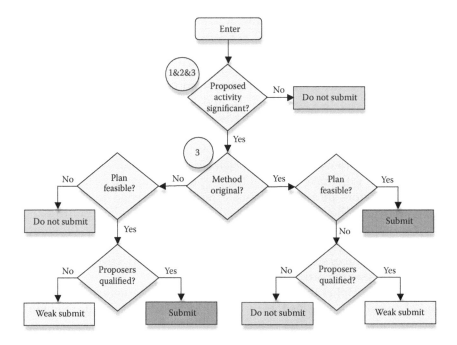

Figure 2.2 Decision diagram for NSF's intellectual merit test. (This decision diagram represents the author's suggestion for testing proposals against NSF's intellectual merit criteria and does not reflect NSF's opinion.)

feasibility should play a more deciding role on the faith of a proposal as shown in the figure. A proposal without a new method and infeasible plan would be a hard sell to most reviewers even if the proposed activity were significant unless the proposers are highly qualified. The qualification of the proposer should be based on the publication record and recognition of the proposer in the field. Submitting a proposal in a new field without first establishing a reasonably strong record of publications is not recommended. One or two conference presentations and preferably one publication in a reputed journal will go a long way toward establishing a stake in the field. The answers to the questions in Figure 2.2 should be provided as objectively as possible to make an objective decision. However, the decision in favor of or against a submission of proposal may ultimately be determined by other factors such as the funding needs of the proposer, potential risks of rejection, and whether the proposal is written by a single proposer or it is collaborative, among others.

2.1.2 The European Research Council's proposal evaluation criteria

The European Research Council's (ERC) grant schemes guide* does not explicitly use the term *intellectual merit,* but in Section 1.3.4 on page 41 of the guide, it is stated that "excellence is the sole criterion for evaluation" and that "it will be applied to the evaluation of both the principal investigator[†] and the research project."

The evaluation criteria for the research proposal are captured with the following questions in the guide:

Significance of Proposed Activity
- To what extent does the proposed research address important challenges at the frontiers of the field(s) addressed?
- To what extent does it have suitably ambitious objectives, which go substantially beyond the current state of the art?

Methodical Effectiveness of Proposed Activity
- To what extent does the possibility of a major breakthrough with an impact beyond a specific research domain/discipline justify any highly novel and/or unconventional methodologies (high-gain/high-risk balance)?

---

* European Research Council, ERC grant schemes guide for applicants for the advanced grant, 2011 call. Version of November 11, 2010.
† Funding agencies commonly use the terms "principal investigator" and "investigator" in their documents. They will also be used in this text interchangeably with the term "researcher."

Feasibility of Proposed Activity
- To what extent is the outlined scientific approach feasible?
- To what extent is the proposed research methodology appropriate to achieve the goals of the project? To what extent are the resources requested necessary and properly justified?

Qualification of Proposers
- To what extent is the principal investigator's record of research, collaborations, project conception, supervision of students and publications groundbreaking and demonstrative of independent creative thinking and the capacity to go significantly beyond the state of the art?
- Is the principal investigator strongly committed to the project and willing to devote a significant amount of time to it?

As in the NSF's intellectual merit test, these criteria can be captured by a flowchart as shown in Figure 2.3. The flowchart reflects that the ERC's evaluation criteria are more rigid and lead to a single path of submission recommendation and another path of weak submission. In both cases, the proposed activity is rated to be significant and the method is

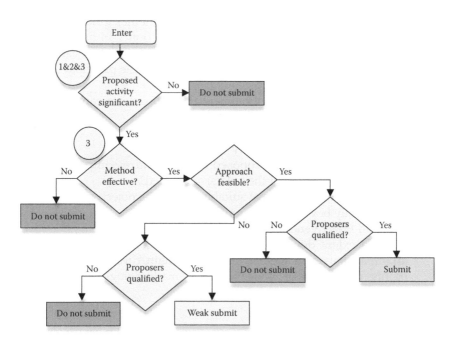

Figure 2.3 Decision diagram for ERC's intellectual merit test. (This decision diagram represents the author's suggestion for testing proposals against ERC's proposal evaluation criteria and does not reflect ERC's opinion.)

rated to be effective. The weak submission recommendation may be too optimistic since the ERC's evaluation criteria emphasize excellence in all aspects of proposal evaluations. Nonetheless, if a proposal only fails the feasibility of approach, it may still be viewed as fundable by a sizeable group of reviewers.

2.2   What is the broader impact and why is it important?

A strict interpretation of the phrase *broader impact* of a research proposal encompasses every possible area of activity in which it finds some use or application. The areas of activity can be:

1. Research activities in other fields in which the results of the proposed research are cited and used.
2. Development and commercialization activities in which the ideas discovered in the proposed research find application.
3. Educational activities, where graduate and undergraduate students may benefit directly from the research project.
4. Outreach activities, where the results of the research project may be disseminated more broadly to enrich the scientific knowledge and experience of the society at large.

We discussed the broader scientific impact of a research proposal in the context of NSF's intellectual merit criteria in the earlier section. The remaining three areas of broader impact can have a significant affect on the success of a research proposal in winning funding. If program solicitations are geared toward the application of ideas, the proposed research should establish a direct link between its expected contributions and their applications to solving practical and technological problems. In some cases, proposals may be required to provide a specific description and plan of work to demonstrate how its scientific findings can actually be applied to solve practical problems or trigger a transition to a new technology. However, the proposed research should not mix up the research goals of the projects with the development needs of a new product or technology. There is always a potential risk of going too far in tying the goals of a proposed activity with a potential technology or product that may result from its research outcomes. For example, in computer systems architecture research, most proposals introduce new design alternatives for microprocessors, memory subsystems, or software layers. The emphasis of such research proposals should be placed on potential system performance improvements rather than implementation and development details of the proposed system. The same criterion applies to other types of systems proposals as well.

The broader impact of a research proposal in educational fields and outreach activities can manifest itself in a variety of ways. In general, there is a trend among researchers to state some general activities when they list the broader impact of their research proposals. These statements include such sentences as:

- The proposed research will train graduate students in an emerging area of research ...
- The proposed research will involve undergraduate students in the research project ...
- The proposed research will train teachers in local secondary schools ...

These statements all point out that the proposed research would like to meet the broader impact criteria with as little effort as possible. They all acknowledge what will inevitably happen when the proposed research engages in educational and outreach activities but nothing specific has been said to make the proposal unique in engaging such activities. As such, they contain no information and will likely make reviewers very skeptical about the intentions of the proposer. What needs to be done instead is to spell out what new ideas will be introduced in educational and outreach activities. For example, if the proposed research should involve secondary school teachers in its proposed activities:

- How will the teachers participate in the project; that is, what will they do?
- Will short courses be developed to disseminate the findings from the project to teachers?
- Will the proposer actively participate in the dissemination of new knowledge to students? If so, how?
- Will teachers and students from different schools meet and discuss the new ideas generated by the proposed research project?
- Will students design and run experiments or build systems to demonstrate the ideas in the proposed research project?
- Will the proposer work with the local public school systems to initiate research training and activities for talented students?
- Will the project results be posted, animated and shared over the Internet?
- Will these efforts affect the curricula of the schools and, if so, how?
- How will the effectiveness of all these educational and outreach efforts be measured?
- Will the experiences and lessons learned from these activities be presented in conferences and published in archival journals?

A cohesive broader impact description for educational goals can be developed for any research proposal by expanding on these ideas and questions. The resulting composition should be specific to the goals of the proposal and its context as much as possible. Whenever possible, such a description should also include prior experience of the proposer in educational and outreach activities. Any preliminary work or plan of work with a particular educational program on or off campus can demonstrate the level of commitment of the proposer to educational and outreach activities, and increase the chances of winning an award.

2.3 Should every proposal have a "real-world application?"

There has been a recent emphasis by funding agencies for research projects to connect their results with the solutions of real-world problems. It will be an exaggeration to state that every proposal can be linked to a practical application as some research projects are very theoretical in nature and it may be a big stretch to tie them with a practical problem. Still, it is highly probable that reviewers would like to see that a proposal connects with the real world in some form or another even when it is extremely theoretical. Several scientific results that were discovered in physics and mathematics in the 19th and 20th centuries were very theoretical but some found applications in real life in the end. An extreme example is the relation between logic circuits and Boolean algebra* that was introduced by the English mathematician George Boole in 1865 (reduced to switching algebra[†] by Claude Shannon). Such circuits process and perform every conceivable computation in today's silicon computer chips. As in this example, some concepts may take quite a while before they find applications but with the rapid pace of changes in today's science and technology fields, funding agencies are more insistent on having proposals that have some relevance to the solution of pressing needs and problems.

Leaving out some of the very pure and most theoretical areas of mathematics, it is not difficult to motivate proposed research by identifying a set of potential applications in most fields of science and engineering. However, as in the broader impact discussion we had earlier, this relation between the proposed research and potential applications must not be generic but rather be very specific. To illustrate this, perhaps with a partially realistic example, suppose that a research proposal seeks to discover a new set of algorithms to perform matrix operations more effectively.

---

* Boole, G. 1854. An investigation of the laws of thought on which are founded the mathematical theories of logic and probabilities. Walton and Maberly.
[†] Shannon, C. 1938. A symbolic analysis of relay and switching circuits. *Trans. Amer. Inst. Elect. Eng.* 57:713–723.

The obvious and very broad application of such algorithms is almost any computation that involves matrices. In theory, it is not impossible for a research project to develop a set of radically different matrix algorithms that may revolutionize computing in general. However, this should be a rare occurrence since science would otherwise advance at a much faster pace with such breakthrough discoveries. Therefore, in most cases, a theoretically motivated proposal must provide a more modest relation between its expected results and the practical problems to which such results can be applied. In this example, the following types of questions may become relevant to narrowing the relation between the anticipated results and their applications:

- What assumptions are made in developing the new matrix algorithms and how do such assumptions affect the applicability of the algorithms?
- What will be the minimum space of computations to which the new algorithms can be applied with a significant advantage over the existing algorithms?
- Do the new algorithms require major changes in the representation of information, numbers, symbols, and so forth, and if so, how complex is it to implement these changes?
- Are the new algorithms as reliable and as accurate as existing algorithms, that is, do they work within the same margins of errors and reliability?
- Are the new algorithms as adaptive as the existing algorithms, that is, can they be as easily transformed between different domains of computation?

As in the broader impact example, this list can be extended to address the potential issues and questions that can be raised by reviewers. Nonetheless, most research proposals focus on a small set of problems to avoid the criticism that they are vague and too broad. Consequently, the potential applications will likely be very specific as well and their description should closely follow the main theme of the proposed research.

2.4   When should a research proposal be submitted?

There is no universal rule as to when a research proposal should be submitted. Funding agencies rarely discourage researchers from submitting proposals. Similarly, researchers find themselves under a lot of pressure by their organizations to submit proposals. This leads to an interesting paradox. If funding agencies receive fewer proposals, then researchers are more likely to win more awards; but this means that researchers should submit fewer proposals, which makes it less likely for them to win more awards. Funding agencies also feel pressured to encourage researchers to submit proposals as receiving fewer proposals indicates a lack of interest in their research programs. This makes it difficult to persuade their sponsors for funding such programs.

Thus, at any point in time, researchers are under a lot of pressure to submit proposals. Consequently, some researchers feel that they can begin with a germ of an idea and develop it to a research project by bombarding funding agencies with proposals. Some even think that funding agencies must pay the bill for their literature survey and openly include it as part of their research plan. This approach is not very likely to be successful, but even if it is, it is not efficient as it wastes a lot of resources for both researchers and funding agencies.

A more cautious approach is to wait until most of the results are obtained and then submit a proposal to finish off the research by working on the remaining and less challenging problems. This is often triggered by the concern that revealing ideas too soon in proposals carry a risk of losing them to other researchers. Even though this sounds like a legitimate concern, there are several mechanisms that are designed to prevent such plagiarism of ideas from occurring too often. In fact, some argue that the best way to avoid losing ideas to other people is to publicize them as early as possible. After all, ideas become important as they are shared with others. Research proposals are a good way to publicize ideas and find sponsors to develop them to their full potential with a few preliminary reports and conference papers.

So, when is the best time to submit a research proposal? The answer is when a preliminary work demonstrates that there is tangible potential for a significant discovery. But how can a researcher know that he or she is on the verge of making a significant new discovery? There is no recipe that can determine the precise timing of a significant discovery. Nonetheless, a number of questions can be raised to help approximate this timing. Three of these questions are:

1. Has the proposer done preliminary work in the field of the proposed activity?
2. Has the proposer been recognized in the field of the proposed activity?
3. Has the proposer been published in the field of the proposed activity?

Each of these questions indirectly probes how close a researcher is to generating significant new results. Having done a preliminary study, the researcher is in a better position to gauge the difficulty of solving

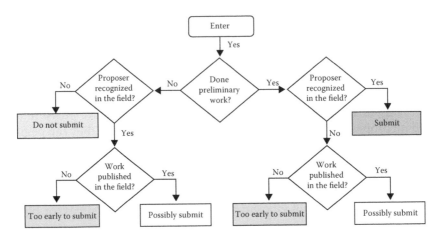

Figure 2.4 Possible timing decision for proposal submission.

remaining open problems. If such a preliminary study also includes a literature survey, then significance of the identified open problems can be better ascertained as well.

It also matters that a researcher be recognized in a field of research to have a reasonable shot at winning a research grant. This recognition is typically measured by the quantity of research presentations in conferences, publications in archival journals, citations to publications, funded proposals, and scientific awards. At the same time, it transcends such quantities, as researchers perceive their fellow researchers by the significance of their contributions more so than the quantities of their output. Researchers rely on different styles of research to gain recognition. Some publish one or two important papers in a field, then move on to other fields. Others continue to work on different aspects of the same research problems within a field. As long as they get recognized and earn their recognition, the style does not matter.

Finally, beyond a preliminary work and broad scientific recognition in a field of research, it is important to actually present and publish the preliminary findings to signal potential reviewers that more results are to follow. The completion of such a phase of activity is an important determinant to judging the timing of submitting a proposal.

Figure 2.4 summarizes the possible decisions under different combinations of these three questions. Obviously, submitting a proposal without any preliminary work and lack of recognition of a proposer's work in the field cannot be recommended. Likewise, a researcher recognized in a field of research who has conducted a preliminary work is ready to submit a proposal. In the remaining four cases, the recommendation is split between too early to submit and possibly submit.

2.5 *Calibration of preliminary research ideas*

The essence of the foregoing discussion is that one should have a credible idea before writing a research proposal. There is a myth that the proposal submission process can help researchers get their ideas across and calibrate them against other ideas. By repeatedly submitting improved versions of a proposal, one may hope to eventually put together a winning proposal. The danger is that the system remembers and recognizes that the proposal has had flaws to begin with. This could potentially risk winning an award and be very damaging to the credibility of a researcher as well. It will thus be helpful if a researcher can predict the potential of a preliminary investigation to lead to a successful proposal and minimize the risk of being rejected without such an engaging and trying process.

One approach to test and calibrate preliminary research ideas is to resort to the evaluation metrics announced in funding solicitations. In the case of the NSF's criteria, such a calibration model can be designed to rely on two metrics: intellectual merit and broader impact. As described in Chapter 1, the intellectual merit captures the scientific integrity and originality of a research proposal. The broader impact captures its potential to influence other research and development projects. Thus, using the scientific integrity, originality, and broader impact, we can assess the potential of a preliminary investigation to lead to a successful proposal. However, although the integrity and originality of a preliminary investigation can be quantified, it is difficult to measure its broader impact, as such an investigation is yet to have a sufficient exposure and publicity in its field.

To help alleviate this problem, we can use another set of metrics that measure the (a) scientific integrity (SCI), (b) affinity to existing research (AER), and (c) separation from existing research (SER). The AER metric measures the degree of rooting that a preliminary investigation finds in an existing field of research. In other words, it determines how much overlap exists between a preliminary investigation and with the core of ideas and problems in an existing field of research. The SER metric measures the degree of separation between the findings of a preliminary investigation and the results known in the field of research in which it is rooted. Here, we assume that every research effort must be rooted in some established field of research even though it may stray far away from it in the end. Thus, the AER and SER can collectively determine how deep a preliminary investigation is rooted in an existing field of research and how far it will be able to move the field from its current frontiers. The scientific integrity of the preliminary investigation provides the support for these two metrics by backing them with the credibility of its methods and claims.

In general, a preliminary investigation that is worthy of submission for funding should perform well in all three metrics. An easy way to visualize this will be to use a radar chart with each metric given a range of

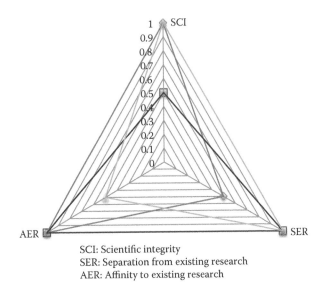

SCI: Scientific integrity
SER: Separation from existing research
AER: Affinity to existing research

Figure **2.5** Characterization of three preliminary investigations with radar charts.

values. Suppose that all three metrics are defined to draw values from the interval [0,1], where 0 represents the worst performance and 1 represents the best performance with respect to each of the metrics. Higher values of SCI may be interpreted as having more provable or verifiable results and claims in the preliminary investigation. Higher values of AER should indicate that preliminary results have a stronger foundation in the field and higher values of SER should indicate that the preliminary results separate further from existing results.

As an example, the radar chart in Figure 2.5 depicts the (SCI, SER, AER) metrics for three preliminary investigations. For each preliminary investigation, one of the three metrics is set at the half point, whereas the other two are set to their maximum values. It is not easy to predict which of these preliminary investigations is more likely to lead to a successful research proposal. A possible approach is to use the area of the triangle formed by joining the points along the three metric axes, as this would measure the confluence of three metrics geometrically. However, in this case, all three triangles have the same area, suggesting that all three preliminary investigations are equally probable to win funding. To differentiate further, we can stipulate that the higher values for SCI and SER may be viewed more favorably as they indicate that there will be credible and new results if the preliminary investigation is expanded to a comprehensive research proposal. A higher value of SCI can partially compensate an apparent weakness that might be indicated by a low AER value. Still, it is important to apply certain thresholds to all three metrics in order not to end up with a weak proposal in any of the three categories. What is important to realize is that a radar chart only presents a graphical view of the raw facts about the characterization metrics for a preliminary investigation and their interplay, and nothing more. Furthermore, specifying metrics to measure the potential of a preliminary investigation is only part of the answer. Researchers should use their own judgments to map the [0,1] interval to their preliminary investigations for each of the three metrics. This may require a dynamic process that may have to be calibrated over time using the realized rate of success of preliminary investigations by the number of citations and funding the investigator receives.

2.6 Is every investigator equally qualified to win funding?

It is well-known that most funding agencies ask investigators to submit a document summarizing their professional accomplishments and demonstrate that they are qualified to carry out the research they propose. Most funding agencies also dictate rules for submissions of proposals that may exclude certain populations of researchers. For example, career awards are restricted to academically young researchers. Other solicitations allow only senior-level researchers with an established track record to submit proposals. For example, ERC solicits proposals in two categories: starting independent researcher grant and advanced investigator grant. The latter grant solicitation requires that senior investigators have 10 or more publications in leading peer-reviewed journals of their respective research fields or some equivalent publications that include patents and conference papers. These general restrictions cannot be avoided, as funding agencies set them, but there are other factors that can potentially place some researchers in a more advantageous position than others. As mentioned earlier, some of these are investigators' professional qualifications, fields of research, track records and funding levels, and their home institutions and research environments, among others.

Ideally, proposals should only be evaluated on their own merits, but reviewers and panels invariably focus on the qualifications and track records of investigators. To give an extreme example, a computer scientist who proposes to investigate the hunting behavior of animals in the Amazon rain forest will be much less likely to receive a favorable decision than an animal scientist would. Even within the same field of research, professional qualifications and research expertise weigh heavily on the minds of panelists and reviewers. Thus, building a strong and credible record in a field of research has a significant impact on the hit rates of investigators in receiving funding for their research proposals. In practical terms, this means publishing significant articles in archival journals and presenting ideas at conferences where fellow researchers congregate and discuss emerging scientific problems in their fields of research.

The home institution and research environment of an investigator also factor in the success of research proposals, especially in experimental research. Proposals that rely heavily on experimental methods cannot succeed without adequate laboratory facilities and their reputations among the research community of the investigator. In the case that an investigator lacks such facilities, working with government or private research centers and laboratories would be very helpful.

2.7   Using online bibliometric tools for self-assessment of research qualifications

A reliable source of self-assessment of one's research qualifications in a field is the performance of the pool of researchers who work in the same field. How many research papers have they published? How many citations have they received? How much funding have their research projects attracted? How many doctoral students have they supervised? How many editorial assignments have they been given? We will examine some of these questions further in subsequent chapters. Here we will illustrate how an online bibliometric tool may be used to assess one's research publication performance.

Suppose that a researcher working in applied physics would like to submit a research proposal on nanotubes, but before doing so, he or she would like to determine if they have sufficient research credentials in the field. One way to accomplish this would be to survey the field of nanotubes and determine the profile of an average active researcher in the field. There are several online bibliometric tools and databases that can be used to accomplish this. Some are more specialized in certain fields of study, whereas others provide more comprehensive coverage of many subjects extending from social sciences to life and physical sciences.*

In our example, the Web of Knowledge^SM provides an ideal tool for gathering all the relevant information and statistics as it has an extensive database for searching research articles in science and engineering.† To begin our session, we first connect to the Web of Knowledge and then

---

* Exactly what tool and database would be best to use will largely be determined by the subscriptions of the home institution of a researcher. Most universities subscribe to Web of Knowledge, Scopus, and Inspec. Many universities also make other more specialized tools and databases available. It is best to check with your library coordinator to find out how you can access them from your computer.

† Web of Knowledge, http://wokinfo.com/.

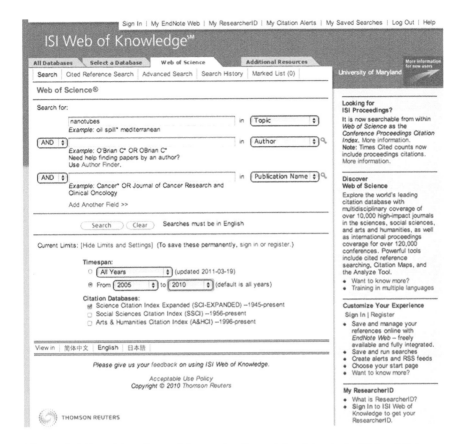

Figure 2.6 Searching for articles under "nanotubes" topic in Web of Science. (Web of Science^{SM}, Thompson Reuters, www.science.thompsonreuters.com. With permission.)

enter "nanotubes" in the topic text field* and limit the search to the years between 2005 and 2010, to ensure that only active researchers are included in the search as shown in Figure 2.6. We also limit our search to the Science Citation Index Expanded^{SM} (SCI-EXPANDED) database to focus on articles that are related to physical sciences. Including the other databases would likely return more articles, but we assume that nanotube research articles should be stored in physical sciences databases. At this point, we click on the search button and this generates a list of articles whose topics include "nanotubes" as a keyword. More precisely, the Web of Knowledge defines *topic search* as a search operation that searches all the words in the search phrase in the union of the title, abstract, author keywords, and key-

---

* We can enter "nanotube OR nanotubes" for a broader search but using "nanotubes" suffices to illustrate the method.

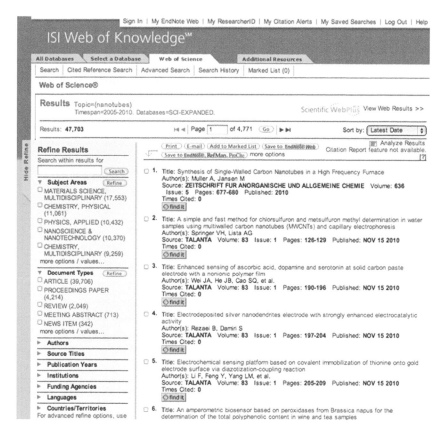

Figure 2.7 Display of results for "nanotubes" topic research in Web of Science. (Web of Science^SM, Thompson Reuters, www.science.thompsonreuters.com. With permission.)

words plus fields. The last field is used by Web of Knowledge to include keywords other than those entered by authors and that may come from references and other places in the article.* The search returns all the articles that contain one or more of the words entered in the search box.† In our case, the phrase "nanotubes" contains only one word and is searched in the fields we mentioned and all 47,703 articles in the database that contain the word "nanotubes" are displayed as shown in Figure 2.7.

---

* Searching for keywords or phrases can be enhanced by logical operations such as AND, OR, NOT, SAME, and parentheses. The description of these operations is outside the focus of this text and can be found at the Web of Knowledge site under a help link.

† It is possible to search for articles whose "topic search" includes multiword phrases exactly as they are entered in a search phrase by enclosing them between double quote marks. This is not needed in our search since "nanotubes" is one word.

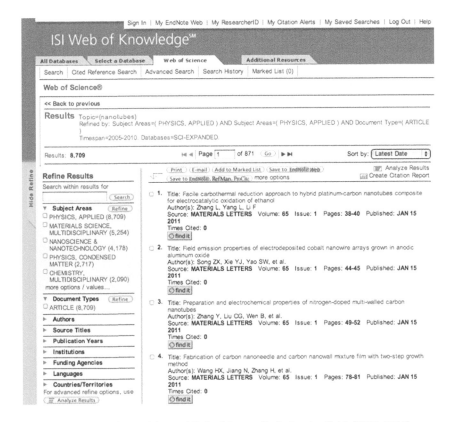

Figure 2.8 Nanotube articles published in applied physics field (2005–2010).

At this point, we can further narrow our search by using the subject area and document type options. Given that our researcher works in applied physics, we click on the "applied physics" and "article" options on the left and then press the refine key. This returns 8709 articles as shown in Figure 2.8, which are listed in descending order of the number of citations they have received. Now that we have narrowed our research, we can generate a citation report for all 8709 articles and check our researcher's record against the various performance metrics listed in the citation report for the pool of papers published in the field. This can be done by clicking the "Create Citation Report" button. This displays the plots and statistics shown in Figure 2.9.

The bar plot on the left shows the number of published articles in the applied physics discipline between 2005 and 2010 whose topic fields include the word "nanotubes." The publication count plot reveals that the number of articles has been increasing since 2005, except for 2010. The plot on the right displays the number of citations received by these articles during the

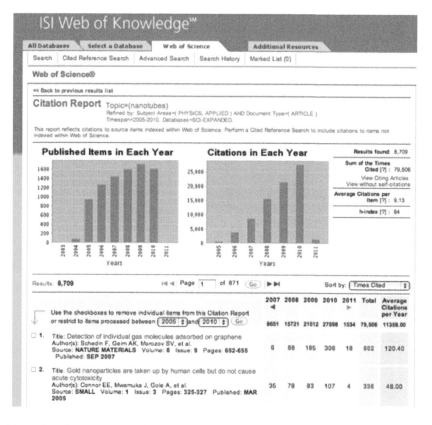

Figure 2.9 Citation report for nanotube articles in applied physics.

same period of time. It shows that the number of citations has increased more than three times since 2007. Both these plots show that nanotubes have been heavily researched in applied physics between 2005 and 2010.

The numerical statistics on the right show that articles meeting our search criteria have received 9.13 citations on the average. This is not the average citation count of the authors of the articles but rather the average number of citations received by an article published between 2005 and 2010. Among the 8709 articles that are included in the analysis, 84 have received at least 84 citations as indicated by the *h*-index in the figure. The highest number of citations received by any of the articles is 602 and the lowest number, which is not shown in the figure, is 0. The distribution of the number of articles receiving between 0 and 10 citations is given in Table 2.1. They were computed by examining the tail of the list of citation counts up to 10 citations. Thus to determine the number of articles with x or fewer citations ($0 \leq x \leq 10$), we simply counted the number of articles from the last page of articles toward the first page until we located the last article that had received x citations.

Table 2.1 Tail Distribution of the Nanotube Articles
for 0 to 10 Citations

Number of Citations Received ≤	Number of Articles	Percent among 8709 Articles
0	1900	22
1	2877	33
2	3657	42
3	4276	49
4	4796	55
5	5230	60
6	5560	64
7	5884	68
8	6133	70
9	6370	73
10	6583	76

These numbers should encourage our researcher, as publishing one article in a reputable applied physics journal with five or more citations would beat more than 60% of all the articles published in the last five years. By publishing an article receiving 10 or more citations, our researcher would beat more than 76% of all the articles. These numbers track typical funding rates at most funding agencies. In general, researchers should strive to match their average citation counts with the funding rates of the programs to which they submit their proposals. In this case, if a funding agency supports applied physics proposals at a 25% rate, then researchers with an average citation count of 9 to 10 should have sufficient recognition in the field to pass the test on the "qualifications of the proposer to carry out the proposed activity." Of course, the higher the average citation count, the more qualified the proposer would be in the eyes of the reviewers.

The Web of Knowledge does not provide a direct display of the average number of citations received by all the articles written by a researcher in a field. However, this can be accomplished using the Web of Knowledge site. Once we complete our search for "nanotubes" and list all the articles, we can click on the "Analyze Results" link to initiate a second search with the "authors" option to list all or a subset of the researchers that have authored or coauthored one or more of the 8709 articles found. This list will include the names of the researchers together with the number of articles they authored or coauthored. By selecting the names of the authors, we can then conduct another citation analysis on the selected author or authors. For example, we used the Web of Knowledge to list the 10 authors with the highest count of authorship in the 8709 articles as shown in Figure 2.10. The page displayed states at the bottom that some 19,726 authors were missing from the list, as we requested to list only the top 10 authors. Thus, some

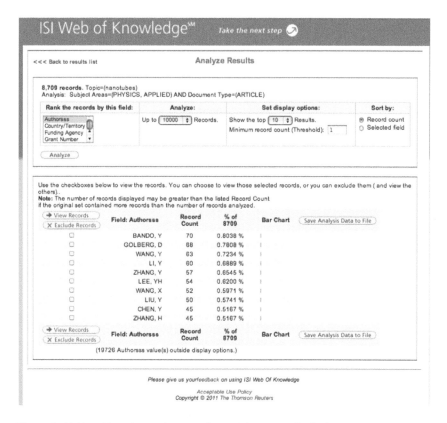

Figure 2.10 Top 10 authors of nanotube articles in applied physics, 2006–2010.

19,726 authors contributed to writing these 8709 articles on nanotubes in the applied physics field between 2005 and 2010. This gives 2.266 authors per article on the average, indicating the degree of collaboration in the field.

We can gather other statistics from this page. For example, if we click on one or more of the selection boxes on the left-hand side of the authors' names and click "View Records" we obtain a list of articles written by the authors along with their citation counts. This new list can be used to obtain the citation statistics of the selected author by clicking on the "Create Citation Report" link as shown in Figure 2.11. This researcher has received 1172 citations from the 8709 articles that were published between 2005 and 2010, had an average citation count of 16.74, and an *h*-index of 22, implying that 22 of the author's 70 articles had at least 22 citations. The citation chart indicates that the number of citations of this researcher has gone from a barely noticeable figure in 2005 to over 300 in 2010 alone. Thus, this researcher should not have any difficulty in communicating his or her ideas in a research proposal. The same process can be used for other

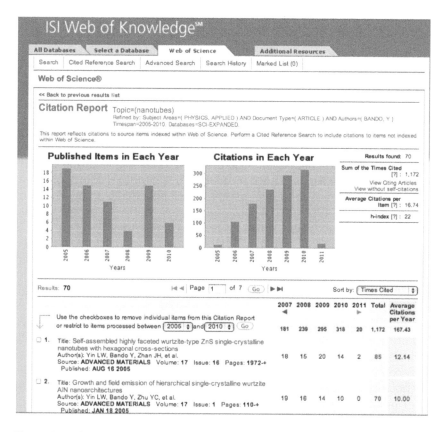

Figure 2.11 Citation record of an author on nanotube articles in applied physics, 2005–2010.

researchers as well as for any subset of researchers of interest to develop a quantitative comparison of our researcher's publication and citation statistics against the competitors in the field.

Obviously all these statistics provide a first-order evaluation, and researchers can supplement them with other criteria to judge if their credentials are competitive for submitting a proposal. In addition, the publication and citation patterns described by the bar charts provide hints about the vitality of the research field. Such bar charts can be used to determine if there is enough potential for the field to accommodate additional funding. In this example, the nanotubes research in applied physics field seems to be very vibrant and should be monitored carefully to see how it develops over the next few years. Again, other statistics that take into account the changes in funding levels of a given field of research can be added to the analysis of the fertility of the field.

2.8   *Collaborating with other researchers*

Collaborations can be beneficial to scientific progress if there is a good match between its participants. Collaborative proposals often bring researchers together with overlapping research specializations. They may also be productive for researchers with complementary research specializations to submit collaborative proposals. Collaborating researchers may be affiliated with the same department or different departments in the same institution or may work at different institutions. In all of these cases, the benefits to all researchers that participate in a collaborative proposal must outweigh the drawbacks of such a collaborative effort. First and foremost, all participants in collaborative research must be equally enthused about the collaboration. Typical collaborations begin with introductory discussions between researchers at meetings and conferences. As in proposals authored by a single researcher, collaborating on a preliminary investigation before proceeding with a collaborative research proposal can help participating researchers to determine if joining forces will be worthwhile. Contributions need not be equal in concept or labor, but each participant must have a clearly defined role for a collaborative research to succeed.

Potential benefits of collaborative research include increased capacity and utilization of available resources, especially in experimental research projects, and scientific interaction and exchange of ideas between potentially competing researchers prior to publication of results. The drawbacks are potential competing interests in the distribution of credits such as ordering of names in the authorships of articles, patents, and other publications; recognition of contributions; allocation of the funds between the collaborators; and the difficulty in decentralized administration and coordination of research activities. Allocation of research problems to graduate students may be particularly contentious as investigators would like to make sure that their graduate students develop original research results in order to complete their programs and graduate. Resolution of intellectual property and patent rights, and sharing of facilities and liability issues in collaborative experimental research may be particularly difficult to overcome when collaborators are affiliated with different institutions. These problems must be addressed, possibly by entering into formal agreements before collaborative research begins to avoid any misunderstanding and conflicts that may encumber the execution of the collaboration.

2.9   *Summary*

This chapter described the key factors that will likely have a direct impact on the probability of proposals winning awards and presented simple tests to determine if a research proposal is ready for submission. Besides the intellectual merit and broader impact, two new metrics, AER and SER, that can be used to assess if a preliminary investigation has reached a stage for triggering the submission of a research proposal were introduced. These metrics are especially valuable to predict the potential broader scientific impact of a preliminary investigation before it is expanded into a full-blown investigation. This chapter also provided step-by-step demonstration of how to use the Web of Knowledge, a widely available online bibliometric resource, to generate and compare publication and citation statistics within a field of research.

2.10   Bibliographical notes

There are other resources on the Web that provide similar search capabilities to Web of Knowledge but their bibliometric and citation analysis techniques are not as integrated to their search engines. Google Scholar™ is one such resource, where articles can be searched by keywords in author, title, publication date, and publication source.* There is a citation counter that is called citations-gadget (a Google Scholar universal gadget for scientific publication citation counting) but it was not linked to the Google Scholar search site at the time of this writing. Science Direct is an alternative Web application with comprehensive coverage of research articles in physical sciences and engineering, life sciences, health sciences, and social sciences and humanities, but it does not include a citation engine.† Scopus is another Web-based scientific document search tool with advanced search functions and a capability to generate citation statistics for authors.‡

2.11   Questions

2.1 Does the flowchart in Figure 2.1 describe the steps you would use to determine the intellectual merit of your proposals? Will you submit a proposal if the answers in the flowchart lead you to conclude a weak submission?

2.2 Discuss the pros and cons of submitting a proposal for funding before you feel that it is bulletproof against the most critical reviewer.

2.3 Compare the decision flowcharts for NSF and ERC proposal evaluation criteria and comment on any similarities and differences you observe.

2.4 In Section 2.2, a number of generic broader impact statements have been given. For example, take the statement: "The proposed

---

* Google Scholar, http://scholar.google.com.
† Science Direct, http://www.sciencedirect.com.
‡ Scopus, http://scopus.com.

research will train graduate students in an emerging area of research ..." How would you add more context to this statement so that it becomes unique to your research proposal and exhibits original ideas of broader impact?

2.5 Do you think that funding agencies should support researchers to conduct preliminary research? Should this support cover investigators' time to do a literature survey and acquire the state-of-the-art concepts and skills in a field?

2.6 Do you think that proposals should be evaluated using a blind review? What would you do if you rate a proposal excellent in a blind review and later find that the proposal was written by a researcher who you consider mediocre in his or her field of research?

2.7 Which of the three metrics introduced in Section 2.4 are more important for you to judge if a preliminary investigation should lead to a research proposal? What threshold value will you use for each metric? Are there other criteria you would use to evaluate the potential of a preliminary investigation to lead to a winning proposal?

2.8 What should investigators do if they cannot publish the results of their preliminary research? Would it be rational for them to extend their preliminary research to a research proposal?

2.9 When you review research proposals, do you reflect on your record of accomplishments? Are you as critical of your own research as you are of others' research?

2.10 Perform a case study similar to the one given in the text in your own field of research and decide if your publication record and recognition in the field warrant submitting a research proposal. Use the publication and citation statistics of articles to predict where your research field is headed during the next five years.

2.11 Which statistics, number of publications, or number of citations are more reliable to predict if the research in a given field is declining? Which are more reliable to predict if it is growing?

2.12 If you determine that the activity in your field of research is declining, would you continue to submit research proposals and work in the field?

2.13 Suppose that you would like to include a section about a possible real-world application of your ideas in a proposal. What metrics will you use to emphasize the potential advantages of such an application over existing technologies?

2.14 What would you do if during a collaborative research, one of your coinvestigators decides to end the collaboration?

2.15 What are most challenging aspects of collaborative research for you? How can you overcome them?

chapter three

Building blocks of a winning proposal

Research proposals play a vital role in the progress of science and provide investigators much needed resources to carry out their research projects. Investigators view writing proposals as both an obstacle and opportunity in the conduct of their research projects. It is not too uncommon to hear them complain that they have to write yet another proposal when they interact at scientific meetings. This concern is not entirely unfounded. It is not simple to articulate new ideas that are not completely ironed out with several potentially open problems in need of solutions. Stating such problems without providing credible ideas to solve them will not likely be sufficient to put together a winning proposal. At the same time, revealing too much information would make the ideas appear to be straightforward, whereas too little information might make them look not as well thought out. Investigators need to put just enough information into their proposals to convince their fellow researchers that they have fundable ideas. There is no single recipe with which this can be done with certainty. However, there are some cardinal facts that should be followed to increase the probability of success. In this chapter, the aim is to (a) explain these facts while describing basic building blocks of research proposals, (b) present some concrete styles that may be used to organize research ideas into a comprehensive proposal, and (c) describe some mistakes that can be avoided when writing proposals.

3.1   How should a research proposal be organized?

One of the most difficult steps in writing a research proposal is to determine where to begin. This is possibly true for any form of writing. Still, putting together a research proposal is particularly more challenging as it involves formalizing open problems and providing credible ideas for their solutions. Perhaps, a relatively safe place to begin is to determine the major sections of the proposal and then, from this skeleton of a proposal, build its contents and arrive at a final document after a few iterations. Fortunately, most proposals have a small set of well-identified sections, which may be called different things by different authors but the functions they serve are more or less the same:

- Summary
- Introduction and background
- Research problems and proposed work
- Approach and preliminary results
- Plan of proposed work
- Expected contributions and impact
- Qualifications of investigator(s)
- References to earlier work
- Access to resources
- Budget and its justification

Each of these sections serves a well-defined function as described next.

The *summary* of a proposal must capture the essence of the proposed work with a concise description of the problems to be explored and approach to be used. It must also state the intellectual merit and broader impact of the proposed research.

The *introduction* must describe the background over which the proposed research will be built. It must survey and discuss the key results that are pertinent to explaining the scientific significance and motivation behind the proposed research and its goals.

The *research problems* section must introduce the specific research problems to be tackled in the proposal. It must then explain what makes these problems worthy of investigation in terms that are understandable by a nonspecialist. Finally, it must formalize them in the language and symbolism of a recognized field of study to allow a specialist (reviewer) to judge the feasibility and complexity of their solutions.

The *approach and preliminary results* section must provide a cohesive scientific framework within which the stated problems can be attacked and potentially solved. It should also demonstrate the efficacy of the approach with some preliminary results—to the extent that this is possible. Proposals with preliminary results, demonstrating that the approach has a good probability of success, are much more likely to receive awards than those submitted without any preliminary findings. This was discussed earlier in Chapter 2, Section 2.4.

The *plan of proposed work* section must describe the organization and layout of the work to be performed as precisely as possible. It should include work packages and a timetable, indicating the approximate dates for the completions of the work packages and expected key accomplishments of the proposal. The work packages may consist of theoretical studies, designing simulation models and experiments, prototyping and testing systems, and making field trips and expeditions. The work plan must also describe the interdependence between different work packages. It must further offer a fail-safe strategy to explain how the proposed research can ensure a reasonable likelihood of success in the event that some of the work packages cannot be completed as initially planned.

The *expected contributions and impact* section must describe the results that are likely to be obtained by the proposed research as accurately as possible. These expected contributions must not be overly ambitious or overly modest. They must permit the reviewer to have a realistic evaluation of their feasibility, significance, and broader scientific impact in the investigator's field of research and adjoining fields.

The *qualifications of investigator(s)* section must demonstrate as succinctly as possible the competence of the investigator(s) to carry out the proposed research. Funding agencies enforce this by allowing the investigators to list a fixed number of publications whose subjects overlap with the topic of the proposed research. For example, the National Science Foundation (NSF) allows each investigator of a proposal to list up to five publications most closely related to the proposal and a second list of up to five significant publications irrespective of their subject relation to the proposal. ERC requires 10 significant publications for its advanced grant applications. If permitted by the funding agency to which the proposal is to be submitted, the investigator's citation record may also be included in this section. In addition to the publication and citation information, the qualifications must also state the professional accomplishments of the investigator. These should include undergraduate and graduate degrees, positions and ranks held at places of employment, and a description of significant synergistic past educational and outreach activities to demonstrate leadership and initiative for broader educational impact. Other requirements include the list of collaborators of the investigator and other members of the research team to eliminate potential conflicts of interest in the selection of reviewers. For example, doctoral dissertation advisors may not serve as reviewers of their advisees' proposals and vice versa. Similarly, researchers may not serve as reviewers on proposals written by those who work at the same institution with them.

The *references* must cite all the articles related to the problems stated in the proposed research including the most recent publications. It should not be forgotten that reviewers are selected from the experts in the field. They may be offended if their work is not cited in the proposal when they think that it should have. When citing results from an article, it is fine to be critical, but one must be discrete and not be subjective. Each citation should be given its due credit by explicitly relating its results with the problems addressed in the proposal. Undeserving citations may appear awkward and make reviewers suspicious of the intentions of the investigator. Besides those articles directly linked with the proposed research, some survey articles or articles that may have a partial overlap with the theme of the proposal should also be cited. This would make reviewers feel comfortable about the investigator's breath and depth of knowledge in the field and familiarity with research results that transcend his or her own immediate focus of research.

The *access to resources* section is more critical in experimental and empirical research proposals. However, even in proposals that have a little or no experimental focus, investigators must demonstrate that any resources including graduate students, technical and nontechnical support staff, and graduate courses offered in their units or institutions meet or exceed what would be needed to carry out the proposed research. For example, if a research project requires hiring graduate students with programming skills in a particular programming language or software development technology, it must be demonstrated that such students are available. Similarly, if an experimental research project requires technical skills to handle or operate special equipment, the availability of such skilled technical personnel or students must be demonstrated.

Finally, the *proposal budget* must be carefully prepared to cover all the direct and indirect expenses that will be incurred during the investigation of the proposed research. The various line items in the budget, including the salaries of the investigator(s), and tuitions and stipends of graduate students must match the specified efforts. Likewise, the resources must be appropriately budgeted to complete all the described work packages in a timely fashion and according to the timeline given in the proposal. Budgets with exceptionally high funding requests are scrutinized more thoroughly by funding agencies and reviewers. Therefore, extreme caution must be used to justify the requested funds. Budget topics will be further covered in Chapter 4.

3.2 *Proposal summary templates and samples*

The summary of a research proposal presents an opportunity for investigators to showcase their new ideas and their significance before reviewers get to read the rest of the proposal. The first impression is always important. As stated in the preceding section, the summary should succinctly describe (a) the proposed work (problem statement and approach), (b) the intellectual merit, and (c) the broader impact of the proposed research. These three parts can be organized in any given order and even mixed together, but the common proposal summary styles are shown in Table 3.1.

The first style can be viewed as a *bottom-up* description of a research proposal. As depicted in Figure 3.1, the problems to be investigated by the proposal and approach are stated first. Next, the scientific integrity of the proposed work and its originality are summarized. Finally, expected contributions or outcomes and their broader impact are described. Here,

Table 3.1 Possible Proposal Summary Styles

Style	Order of Presentation		
	Bottom	Middle	Top
Bottom-up	Proposed work	Intellectual merit	Broader impact
Middle-first-top-last	Intellectual merit	Proposed work	Broader impact
Top-down	Broader impact	Intellectual merit	Proposed work

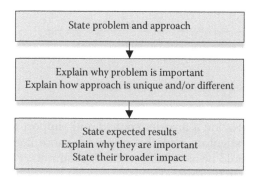

Figure 3.1 Bottom-up proposal summary style.

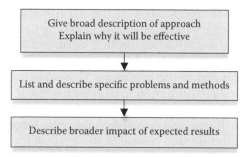

Figure 3.2 Middle-first-top-last proposal summary style.

the transition from the statement of the proposed work to the description of the intellectual merit can be accomplished by emphasizing the originality of the problems and approach. Similarly, expected contributions can be used to transition from the intellectual merit statement to the broader impact part. This provides a natural flow in a project summary and is highly recommended for most research proposals.

In the second style, to be referred to as the *middle-first-top-last* approach, the description of intellectual merit is moved to the beginning of the summary and the broader impact is kept at the end as shown in Figure 3.2.

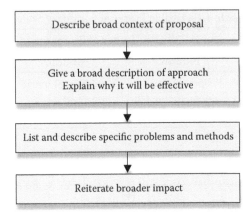

Figure 3.3 Top-down proposal summary style.

This is done to provide a broader context and support for the proposed work. Some investigators find this approach more natural for writing summaries. It should be as reliable as the first summary style and can be used in most research proposals.

The third style combines the three parts of a proposal summary in reverse, and can be viewed as a *top-down* approach as shown in Figure 3.3. The intellectual merit bonds the two aspects of the proposal in the middle to demonstrate that the proposed work has a reasonable chance of accomplishing the broader scientific impact declared at the beginning of the summary. The proposed work, intellectual merit, and broader impact sections are rarely delimited in this style. In some cases, the broader impact is reinforced at the end of the summary. This style of summary works best when there is a pressing need to deal with a major crisis or disaster. Reviewers will likely be sensitized to proposals that can help deal with such emergencies. It can also be used to describe large-scale proposals where the big picture is more central to the goals of the project than the subproblems and methods used to tackle them.

It will be helpful to examine a few research proposal summaries published by NSF to get a better understanding of how these different styles can be applied to writing actual proposal summaries. We begin with the bottom-up summary style.

3.2.1 Example: Bottom-up summary style

An example summary matching the bottom-up style appears next and describes a proposal funded by the Chemical, Bioengineering, Environmental, and Transport Systems (CBET) Division at NSF. The elongated text describes the actual summary as it appears in the NSF cite.

The summary begins with the description of the proposed work. The investigator would like to program two smart functions into biomaterials and evaluate these functions in the form of an injectable vascular patch. He asserts that the oxidative stress response method is a good fit for this investigation as it is used as a universal mark of damaged organs, tissues, and cells. The rest of the proposed work section describes the approach with specific terms that can be deciphered only by a specialist in the field, as indicated by the italicized phrases.

Career: Oxidative stress—Responsive
shape memory vascular patch*

Proposed work (inserted in the text): *The PI will program two "smart" functions* (i.e., oxidative stress-responsive and shape memory functions) into bio-materials by employing shape memory polymers cross-linked with peptide sequences that degrade in response to injury-mediated overproduction of reactive oxygen species. *The "smart" functions of these biomaterials will be evaluated* as a form of an injectable vascular patch whose shape, size, and thickness can be tuned to custom-fit even small blood vessels. Stimuli-sensitive materials change their structure and shape in response to changes in environment, such as heat, light, moisture or magnetic field. *In the proposed project, oxidative stress is selected* as an external stimulus because overproduction of reactive oxygen species is a universal mark of damaged organs, tissues, and cells. *Reactive oxygen species-degradable peptides* that can cross-link polymers when making patch scaffolds *will be used* to generate the oxidative stress-responsive function. Shape memory polymers in a combinatorial format, x% cross-linkable unit -co-y% non-cross-linkable unit, where x and y% indicate the molar ratio, will be used to tune the ability to memorize temporary shapes and regain their original shape after exposure to body temperature (37°C). The basic polymer type proposed for this study has shown excellent vascular compatibility in previous

---

* Sung, H.-J. 2011. Career: Oxidative stress—Responsive shape memory vascular patch. NSF Career Award No. 1056046, http://www.nsf.gov/awardsearch/showAward. do?AwardNumber=1056046.

studies. Furthermore, the *polymer-peptide complex will be fabricated* as vascular patch scaffolds aimed at repairing small blood vessels (e.g., cerebral hemorrhage and stroke). The *patch scaffold surface will be coated* with antibodies against vascular cell adhesion molecule-1 to generate a suture-free sealing effect by mimicking inflammatory cell adhesion onto injured endothelium. The smart functions and their *subsequent effects on vascular healing will be evaluated* in a bioreactor system that mimics the vascular environment. This is a very challenging research task due to the interdisciplinary nature of the conceptual and technical approaches.

The last sentence in the proposed work section is used to transition to the intellectual merit statement. It indicates that the proposed work presents a challenging research due to the interdisciplinary nature of its conceptual and technical ideas. Thus, the potential reviewers of the proposal are gently told that the proposed work has intellectual merit. This is reinforced in the intellectual merit section given next. The italicized phrases reflect the investigator's confidence about the originality of the problems on which he proposes to work. He also demonstrates the scientific integrity of his ideas in the last sentence: The proposed work requires a "deep understanding of material design and fabrication" process.

Intellectual merit: The proposed research *will advance* the state of the art in methods and techniques for applications of "smart" biomaterials to develop therapeutic inventions. *Some of the innovative expected results include*: (1) *providing a stepping stone to develop the next generation of biomaterials* that enable artificial intelligence-like work flow (i.e., navigating, sensing, and fixing); (2) *incorporating biological molecules into a shape memory material function*; (3) *advancing "smart" material functions to cope with complex biological signaling*; (4) *a new therapeutic approach* to regeneration of injured small blood vessels and a therapy of further pathogenesis, such as cerebral hemorrhage and stroke; and (5) *generating a new tool box* for minimally-invasive surgery and for design of scaffolds with customizable size, shape, and thickness. *The proposed research will have a far reaching impact in terms of real applications of the research developed, as well as the broad spectrum of research areas.*

> *Achieving the goals of this project requires a deep under-standing of material design and fabrication, as well as biomedical applications.*

Finally, the investigator tackles the broader impact question by first out-lining the broader scientific impact of the proposed research (first two sentences) and then describing its educational benefits in terms of fairly detailed and original ideas.

> **Broader impact:** *Direct outcomes of this research will influence a large cross-section of the engineering and bio-medical community and will stimulate education towards multidisciplinary subjects.* Successful translation of the research to real world applications *has the potential to revolutionize* biomaterials and the regenerative med-icine industry. *Broader impacts also result from a range of education and dissemination activities,* including (1) integrating research projects with coursework, out-reach, and training through the existing courses and programs; (2) teaching abroad to expand out-reach; and (3) creating a Web-based vibrant youth community comprised of students in the United States and S. Korea. *Though the focus is biomaterials and tissue engineering, the basic elements will appeal to students from all areas of science and engineering. The PI will use established assessment tools to evaluate and adjust his pedagogical methods in these various integra-tive learning environments. He will also disseminate his new pedagogical methods for use by other instructors.*

This proposal summary is nearly a perfect example of a bottom-up proposal summary style.

The following proposal summary is another example of the bot-tom-up summary style, where the italicized text highlights the key statements in each section. The investigator uses the word "objective" to describe the proposed research. In the intellectual merit section, the investigator states the originality of the proposed work rather implicitly, using the verb "introduces" in the first sentence, and subsequently with the help of phrases "design and explore new DBS strategies" and "safer and potentially more therapeutic." There is also a touch of broader impact in this section. In fact, the last sentence fits the broader impact section more closely and seems to serve as a transitional sentence to the next sec-tion. An expert reviewer can connect the dots and credit the investigator appropriately for broader impact in this case.

Career: Modeling and control of neuronal networks*

Objective: The proposed CAREER program *involves modeling and control of neuronal networks in the brain and its application to the treatment of Parkinson's disease* (PD) using deep brain stimulation (DBS). Specific objectives are to: (i) construct a general approach for modeling complex neuronal networks where interactions occur between different brain nuclei, (ii) design computationally efficient control strategies for such networks, and (iii) apply these methodologies to the problem of restoring pathological network dynamics arising from PD with DBS. Success of the program will fundamentally impact the interface between control systems and neuroscience and create new opportunities for medical treatment of neurological disorders.

Intellectual merit: *The work introduces a systems theoretic framework* to (i) understand electrophysiology and pathophysiology of neuronal networks and the impact of DBS control, and (ii) design and explore new DBS strategies to treat PD with extensive simulation before experimenting on animals and translating to patients. *The modeling framework will eventually lead to safer and potentially more therapeutic* DBS for PD. *The ultimate vision is to apply this transformative methodology* to other neurological disorders (Dystonia, Depression, Obsessive Compulsive Disorders), impacting millions of patients worldwide.

Broader impact: *The proposed plan will also impact education at all levels.* It includes *developing a new course* for senior undergraduate and graduate students on modeling and DBS control of neural systems, and reaching out to *women of all ages with unique workshops* and internship opportunities. Finally, *engineering students will be given the opportunity* to perform electrophysiological experiments with collaborator at the Cleveland Clinic.

---

* Sarma, S. 2011. Modeling and control of neuronal networks. NSF Career Award No.1055560, http://www.nsf.gov/awardsearch/showAward.do?AwardNumber=1055560.

3.2.2   Example: Middle-first-top-last summary style

The next example illustrates a proposal summary that fits the middle-first-top-last summary style. As stated earlier, placing the intellectual merit at the top of the summary is motivated by the desire to provide a context to the proposed work. This is particularly helpful if the proposed work is too focused on a set of problems within a narrow subject of research. This may turn off reviewers who are not too familiar with the proposal's field of research. Investigators use this approach to avoid a potential rejection of their ideas by reviewers. In a way, this style can be viewed as a more cautious version of bottom-up summary style. It should be as effective provided that the transition from the intellectual merit to the description of proposed work is well connected as the following proposal summary demonstrates.

> *All inorganic plasmon-enhanced*
> *photovoltaics with eutectic composition**
>
> **Intellectual merit:** Most silicon-based photovoltaic devices require pure crystalline silicon as the base material, which is costly to produce through current crystal growth processes. *Crystal growth of lamellar heterojunctions* via eutectic solidification of earth-abundant materials (e.g. silicon, magnesium, iron) *has the potential to circumvent the costly crystallization processes* used to purify the silicon feedstock, and allow for enhanced minority carrier collection. *Ultimately, this will result in low-cost solar cells based on earth-abundant materials.* In addition, the manufacture of bulk nanostructured crystals by eutectic solidification can be accomplished at a scale commensurate with that of pure crystalline silicon used in current photovoltaic devices. [The proposed research will focus on the growth and characterization of low cost, and high efficiency plasmon-enhanced heterojunction solar cells through eutectic solidification and block copolymer nanolithography.] Eutectic solidification causes the self assembly of lamellar or rod-like domains with length scales from hundreds of nanometers to micrometers, which are ideal for the efficient

---

* Boukal, A., and Tuteja, A. 2011. All inorganic plasmon-enhanced photovoltaics with eutectic composition. NSF Career Award No. 0954267, http://www.nsf.gov/awardsearch/showAward.do?AwardNumber=1066447.

> extraction of minority carriers in metallurgical grade (impure) materials. *To date, no inorganic solar cells have been constructed with eutectic composition. It is expected that* earth abundant, metallurgical grade materials with eutectic composition, combined with plasmon-enhanced optical absorption, *could possibly lead to the development of a new class of low-cost and high efficiency thin film solar cells.*

The intellectual merit section begins by asserting that the eutectic solidification of silicon, magnesium, and similar materials will likely circumvent the costly crystallization processes, and ultimately lead to low cost solar cells. It argues that this approach can be as effective in manufacturing bulk nanostructured crystals as other known techniques. The scientific integrity is captured in the sentence that begins with "Eutectic solidification causes …" toward the end. The originality is asserted with the statement: "To date, no inorganic solar cells have been constructed with eutectic composition." The investigator has used only one sentence, which is shown between brackets, to interject the proposal's research focus. This is almost essential to spell out the intellectual merit of the proposed research as it is done in rest of this section.

The summary continues with the proposed work given next. The italicized text shows that the investigators have thought through their ideas carefully and designed a sequence of steps that captures the key scientific challenges and novelties outlined in the intellectual merit section.

> **Proposed work (inserted in the text):** *The proposed research will study* the controlled growth and electrical doping of bulk crystals of impure silicide; silicon heterojunctions with nanostructured eutectic composition in an induction furnace, *and characterize* the nanostructured crystal structure and minority carrier diffusion length of these materials. *The nanostructured eutectic materials will then be integrated into working solar cell devices.* The effects of nanostructured eutectic material composition and lamellar heterojunction spacing on solar cell efficiency *will be studied to gain fundamental understanding of the device performance. Plasmonic materials will also be incorporated* into these devices to enhance light absorption, leading to potentially higher solar energy conversion efficiency. These plasmonic materials include highly ordered arrays of silver nanoparticles generated by block copolymer nanolithography.

A standard broader educational impact statement follows the intellectual merit section. Here, the investigators describe how their work will involve local secondary schools in their projects with specific references to local schools and programs as indicated by the highlighted text. These educational impact statements will be more credible if they are elaborated in the proposal.

> **Broader impacts:** The proposed education and outreach activities seek to increase the numbers of underrepresented minorities to enter science, technology, engineering, and mathematics (STEM) disciplines. Several hands-on demonstrations, featuring batteries and solar cells, will be made available to students at *Cass Technical High School in downtown Detroit, Michigan. Two top-performing high school students* will be selected to participate in the proposed research during the summer. Outreach efforts also include participation in a five-week, *Detroit Area Pre-College Engineering Program* that encourages junior high students to pursue careers in the sciences.

3.2.3   Example: Top-down summary style

The following proposal summary partially fits the top-down summary style and was funded by the Division of Environmental Biology at NSF. The investigator did not mark the three sections of the proposal summary but they are relatively easy to detect as they have been delimited with boldface subheadings. As described in Section 3.1, the summary begins with a statement that suggests the proposed work will have a broader impact by addressing the effect of climate warming trends around the globe on trees and forests. In other words, the broader impact is what motivates the proposed study.* This is stressed at the beginning of the proposal so that reviewers are sensitized to the broader impact of the proposed study before the rest of the summary is described. Once this is done, the investigator goes on to establish the intellectual merit of the proposed work by asserting that the forest trees rely on their associations with microbial soil species for their healthy growth, and that little is known about the distribution of such species across the American continent and how they

---

* Of course, this does not rule out other indirect broader impacts. However, at the end of the proposal summary, the investigator reiterates that the proposed research seeks to mitigate the undesirable effects of global warming and views this as its broader impact besides the educational impact.

respond to climate changes. The rest of the summary describes the proposed work, educational impact, and reiterates the broader impact.

> *Dimensions: Functional, genetic, and taxonomic diversity of plant-fungal interactions along climatic gradients and their role in climate change driven species migrations**
>
> **Broader impact (inserted in the text):** Climates are currently warming at unprecedented rates. Species must move to track the changing climate or evolve to tolerate warmer conditions; those that fail to do so face extinction.
>
> **Intellectual merit (inserted in the text):** Most plant species, especially forest trees, rely on intimate associations with microbial species living in soil in order to capture the resources they need for proper growth. Little is known about how these invisible, but very important, soil microbes are distributed across the continent, and how they will respond to climate change.
>
> **Proposed work (inserted in the text):** In this project, the genetic, taxonomic, and functional biodiversity of soil microbial communities from forests across the eastern United States *will be characterized* to test for parallel latitudinal patterns with respect to climate. Trees and soil microbes will likely not move at equal speeds as climates change. Therefore, *experiments will test* the functional consequences for tree growth for situations where microbial species migrate slower or faster than trees. This research will allow for more precise predictions about how forests will change as a function of the changing biodiversity of the fungal symbiont community during climate warming.
>
> **Educational impact (inserted in the text):** This research will have several broader impacts for

---

* Lankau, R. 2010. Dimensions: Functional, genetic, and taxonomic diversity of plant-fungal interactions along climatic gradients and their role in climate change driven species migrations. NSF Award No. 1045977, http://www.nsf.gov/awardsearch/showAward. do?AwardNumber=1045977.

society at large. The project will incorporate citizen scientists from across the eastern US in collecting samples for the soil survey, in order to connect the public with original research. Undergraduate students (including women and underrepresented minorities) will be mentored in hands-on research experiences during the project and encouraged to develop independent research projects.

Broader impact reiterated (inserted in the text): By helping to improve our ability to predict ecological changes in response to climate change, this research will increase our ability to mitigate and adapt to the undesirable consequences of a rapidly warming world.

Another example of the top-down proposal summary style with annotation of the three key sections in boldface style is given next.

*Detecting local earthquakes in a noisy continental margin environment**

Broader impact (inserted in the text): Assessing earthquake risk due to seismicity along the Cascadia margin from northern California to southern British Columbia *is a matter of great public interest.*

Intellectual merit (inserted in the text): Studies of regional seismicity recorded by arrays of seismographs are a primary tool for this purpose, but to date *such studies have been largely limited* to onshore arrays. In the upcoming Cascadia project, onshore instrumentation will be complemented by deployments of 60 or more Ocean Bottom Seismographs (OBS) off the Cascadia coast for several years. A modest deployment of OBSs off the Oregon coast in 2007–2009 *has demonstrated the great difficulty* of separating relevant seismic events in OBS data from impulsive signals of probable biological origin.

---

* Trehu, A., 2011. Detecting local earthquakes in a noisy continental margin environment. http://www.nsf.gov/awardsearch/showAward.do?AwardNumber=1049682.

Proposed work (inserted in the text): *This project seeks to develop* computer automated methods for separating seismic signals from extraneous signals in the OBS data, particularly for instruments located in shallow water near the coast.

Broader impact reiterated (inserted in the text): The project has a number of broader impacts, but by far the most significant is the very high *societal relevance* of developing these techniques for studies of seismicity and seismic hazards in general, and for the Cascadia project in particular.

Reflecting on these proposal summaries, we see that while they vary from one proposal to another, they are all cognizant of the three key ingredients that make up a proposal summary: (a) problem definition and method, (b) intellectual merit, and (c) broader impact. Investigators may check the NSF award site to see what summary styles their fellow investigators use. Beyond that, the exact order in which these three attributes are composed is more of an individual style, and it is best left to each investigator to summarize the proposal's salient features as effectively as possible.

3.3   Commonly used verbs in proposal summaries

In proposal summaries, investigators use certain verbs more often to communicate their ideas. Table 3.2 lists some of these verbs, their frequencies, and potential uses in the problem definition, approach, intellectual merit, and broader impact statements. The frequencies were computed by searching the NSF's award database for active awards. Of the 30 verbs searched, only 6 are used less than 3000 times. Other words of interest include "best" that occurred 2093 times; "excellent," 1465 times; "outstanding," 665 times; "breakthrough," 523 times; "superior," 399 times; "argue," 261 times; "feel," 252 times; "worst," 190 times; "mystery," 140 times; "theorize," 43 times; "incredible," 32 times; "speculate," 29 times; "supernatural," 4 times; "one of a kind," 4 times; "metaphysical," 3 times; and "magical" only once. It is seen that the frequency of a word drops as it strays away from scientific notions or becomes more contentious. It would be wise not to use those words with negative or unscientific connotations.

3.4   How to organize and prose an effective introduction

The introduction section is possibly the most complex part of a research proposal to write since it serves as an articulation point for the rest of the

Table 3.2 Verbs Frequently Used in Proposal Summaries

Verb	Problem	Approach	Intellectual Merit	Broader Impact	Frequency
Analyze	X	X			3000+
Apply		X		X	3000+
Characterize	X	X	X		3000+
Construct		X	X		2402
Decrease	X	X	X	X	872
Demonstrate		X	X	X	3000+
Design		X	X		3000+
Determine		X	X		3000+
Develop		X		X	3000+
Expand		X			3000+
Explain		X	X	X	1452
Explore	X				3000+
Extend		X			3000+
Facilitate			X	X	3000+
Focus	X				3000+
Help				X	3000+
Increase	X	X	X	X	3000+
Integrate		X	X	X	3000+
Investigate	X				3000+
Occur	X	X	X	X	3000+
Predict		X	X	X	3000+
Prove		X			1579
Provide		X	X	X	3000+
Show		X			3000+
Simplify		X			576
Solve		X			2390
Study	X	X			3000+
Understand	X	X			3000+
Use		X		X	3000+
Utilize		X		X	3000+

Note: The list was compiled by searching each verb in active awards at NSF's Award
Search site (http://www.nsf.gov/awardsearch/) on February 22, 2011.

proposal. Without a deliberate approach, writing an introduction may turn
into a black hole and suck all the energy and optimism out of an investigator from completing the proposal. To avoid this, it is helpful to view
the introduction section in a research proposal as a place where the three
key parts of a proposal summary are integrated together. In this approach,
the intellectual merit and broader impact parts serve as descriptors for

the proposed work part. The introduction section is formed by dividing the proposed work into a skeleton of parts, and by applying the intellectual merit and broader impact descriptors to those parts, wherever applicable.

One such skeleton and possible applications of intellectual merit and broader impact descriptors are shown in Figure 3.4. The introduction section is decomposed into seven parts. These parts need not be subdivided further and can be viewed as one or two paragraph descriptions. The parts are displayed in the order they should appear in the introduction. Of course, other skeletons are possible depending on the subject and type of the proposal. For example, experimental proposals may include a part explaining the laboratory facilities; collaborative proposals may include a part describing the synergy between the coinvestigators, unique contributions each coinvestigator would bring to the proposal.

The part for the context of the proposed research should set the stage for writing remaining parts by succinctly describing the field of research within which the key contributions of the proposed research will be accomplished. This should be followed by a description of the widely recognized open problems in the field and important known results that relate to their solutions. Once this is accomplished, the goals-of-research part should introduce the exact set of problems that will be tackled in the proposed research. These problems may be generated from those that have been mentioned as open problems or they can be new research problems. Once the problems are specified, the introduction can transition to the approach part, and once the approach is outlined, the expected contributions of the proposed research can be stated. With all these parts completed, the introduction can be concluded by an overview of the investigator's qualifications to conduct the proposed research. The depth of detail in describing each part should be fixed to control the desired amount of information to be revealed, and by keeping in mind that these preliminary ideas will be further expanded in subsequent sections of the proposal.

As parts of the introduction section are written, their intellectual merit and broader impact attributes can be filled in to begin beefing up the introduction and characterizing the role that these two descriptors would play in accomplishing the objectives of the proposal. As indicated by the arrows on the left in Figure 3.4, the parts for research problems, research goals, related results, approach, and qualifications of the investigator can all be enhanced with intellectual merit remarks. For example, the introduction of new problems in the research problems section would add to the intellectual merit of the proposal. During the discussion of research goals, the significance of the research problems to be investigated in the proposed research can be demonstrated. During the discussion of any results that relate to these research problems, the drawbacks or deficiencies of such results can be pointed out. During the discussion of the proposal's approach, its scientific integrity can be established by a

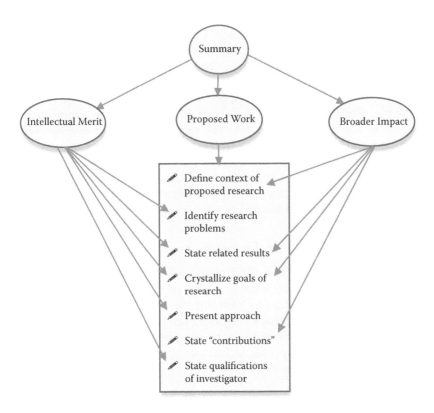

Figure 3.4 Possible organization of the introduction section.

scientific formulation of the introduced research problems and effectiveness of the proposed approach to solve them. Finally, during the statement of the investigator's qualifications, the research experience of the investigator can be used to support successful conduct of the proposed research. Similarly, the context of proposed research, related results, research goals, and expected contributions can all be enhanced by broader impact remarks.

3.5   *Research problems and proposed work*

Just as the introduction section can be formed from the summary of the proposal, this section can be developed by expanding on the preliminary description of the research problems and proposed work given in the introduction. This expansion can be carried out by elaborating on the complexity of the research problems to be explored in the proposed work, and their relation to previously obtained results on these problems.

In some fields of research, such as mathematics and computer science, research problems can be structured and cast as mathematical statements to be proved or algorithms to be written. This facilitates a precise picture of the proposed work to the reviewers. In other fields, informal descriptions may be more appropriate and preferred to state the research problems. However, it is still important to use as precise a language as possible to avoid any ambiguities. The proposed work should be clearly described and its distinguishing traits should be clearly explained. This section often serves as a make-or-break part of the proposal as reviewers will carefully parse the statements in this section to look for the significance and originality of the research problems described, and scientific integrity of the proposed work.

3.6   Approach and preliminary results

As in the research problems and proposed work section, the material in the approach and preliminary results section can build upon what is stated in the introduction to flesh out the main ideas of the approach and present some preliminary results to provide credibility to the ideas described. The description of the approach should go beyond stating the obvious and delve into specific techniques that will be employed to deal with the issues presented by the research problems described in the earlier sections. The complexity or efficacy of the approach as may be measured by the amount of work, time, and other resources should be clearly discussed. Any apparent vulnerabilities or weaknesses of the approach must be carefully analyzed and factored into the decision for advocating the approach.

The importance of having preliminary results prior to submitting a research proposal cannot be overstated, as discussed in Section 2.4. Proposals that include preliminary results exhibit a more cohesive flow of presentation and a better prediction of what can be accomplished in a proposed research. This is only natural since some ground is broken to know where things may go from there.

For avid researchers, conducting research is a continuous process and not having funds temporarily does not mean that they should stop working on research projects and publishing their results. It is only an issue of timing to know when to apply for research proposals. When researchers find that their own time and resources are inadequate for conducting a research project alone, it must be time to write a proposal and seek funding with some preliminary results at hand.

Preliminary results should be provided in sufficient detail but also in a way that should not suggest to the reviewers that the work is already done and that the proposed research can only make marginal improvements on already established results. This does not mean that the work already done by the investigator should be concealed or hidden. The investigator's published results should be cited and included in this section to the extent it relates to the proposed work. Doing otherwise is not only ethically wrong but will also create a bad impression on reviewers and will likely lead to the rejection of the proposal.

It is also critical that the preliminary results be credible. The best way to enforce this is to include only those results that have been published or accepted for publication. Speculative claims and results that are unleashed for the first time in a proposal may lead to criticism and suspicion. This would hurt the proposal's chances in receiving a favorable decision. What is expected in this section at a minimum is a new method or a new application of an existing method that is demonstrated to work by a set of preliminary results and has the potential to produce more significant results in the investigator's field of research and possibly in adjoining fields.

3.7   Plan of proposed work

The plan of proposed work section is where the proposed research takes its most concrete form. In this section, the research problems and approach should be connected with the resources requested in the budget. A clear timeline should be given for the delivery of expected contributions. In addition, the workloads of the investigator and other researchers need to be stated as precisely as possible.

At a minimum, a work plan should spell out the schedule of the work to be performed during the execution of a research project. It should describe what will be delivered by the proposed research and when they will be delivered. The deliverables should be interconnected both chronologically and logically to demonstrate that the work plan is more than a mere list of things to do and is designed to obtain the expected results effectively and with a high probability of success. The work plan should also include a contingency plan that can be triggered as a backup strategy to complete the research project if certain parts of the work plan fail to materialize. However, the contingencies should not replace or dominate the work plan since this may suggest that the work plan is too tentative and not well thought out.

One way to accomplish all these tasks is to use a timetable, where the requested duration of the proposed research is divided into equal segments of time, which are appropriate to characterize the significant and

Table 3.3 Sample Work Plan

	Timeline (in 4-month Increments)								
Task	1	2	3	4	5	6	7	8	9
Work package-1	PI	PI							
Work package-2		PI, RA-1	PI, RA-1	PI, RA-1	RS	RS			
Work package-3					UA-1	UA-1	UA-1	UA-1	UA-1
Work package-4			PI	PI	PI				
Work package-5				PI, RA-2	PI, RA-1	PI, RA-1	RA-1	RA-1	RA-1
Work package-6					PI	PI, PS	PI, PS		
Work package-7					UA-2	UA-2	UA-2	UA-2	UA-2
Work package-8				PI	PI, RA-2	PI, RA-2	PI, RA-2	PI, RA-2	PI, RA-2
Work package-9					PI	PI	PI	PI, RS	PI, RS

expected milestones. This timetable is used to allocate the available manpower and other resources to work packages to carry out the proposed work. Work packages divide the proposed work into manageable pieces and characterize the key tasks to be performed.

An example work plan is shown in Table 3.3. The timetable is designed by assuming that the proposed work will last for 36 months. The total project time is divided into nine 4-month work periods. Each cell in the table represents an effort by one or more of the researchers who will be involved in the proposed work. In this example, there is one (principal) investigator (PI), one postdoctoral research scientist (RS), two graduate research assistants (RAs), and two undergraduate research assistants (UAs). For example, the cell at the intersection of work package-2 and time division 3 indicates that the PI and one of the RAs will be involved in that work package during that time period. Other scenarios are clearly possible depending on the duration of the proposed work and the estimated duration of the smallest work package that must be handled by the PI and the rest of the research team. If the proposed research involves more than one investigator, they can be added by identifying them with their initials. The 4-month divisions in the timeline implicitly determine the timing of the execution of the work packages. If this must be emphasized further, the numerical indices can be replaced by dates such as 01/11–04/11, 05/11–08/11, and so on.

The work packages should be described in terms of the proposed work and research problems and must be carefully crafted to divide the proposed work into interrelated tasks for maximum efficiency and utilization of the available resources including time. Besides actual research tasks, work packages must also include publication and presentation activities of the investigators and other members of the research team. If further divisions of time are needed to adjust the amount of time to some of the work packages, the cells at the intersections of those work packages and corresponding time periods can be subdivided. Color coding may be used to emphasize relations between the work packages and the timing of their executions. Work packages and work plans will be discussed in more detail in Chapter 4.

3.8   Expected contributions and broader impact

This section serves to end the proposal on a high note after laying out the research problems, approach, and proposed work. The contributions can be linked to the work packages developed in the work plan. The broader impact of expected results on the PI's field of research and other fields should also be described in this section. This section can also include broader impact, resulting from educational and other outreach activities, if it is required by the funding agency to which the proposal is submitted. Some funding agencies may request more than just a sentence to describe such activities, in which case, the investigator may have to develop a detailed plan to carry out such activities and include it in the proposed work.

3.9   Qualifications of researchers

There is no unique style when it comes to listing professional qualifications of researchers. Funding agencies require that certain types of information be included in researchers' biographies when submitting proposals. These often include the place of employment and professional affiliations of principal investigator and co-investigators; listing of the PhD advisor; the names of people with whom the investigator(s) have performed joint work, such as collaboration on a funded research proposal, joint authorships of journal articles, and conference papers. It is important to meet the guidelines of funding agencies in generating this section of the proposal.

A template for a sample biography of an imaginary investigator is shown and can be easily modified to meet the specific requirements of any funding agency.

BIOGRAPHY
Name: John Greenvalley
Title: Professor
Affiliation: Electrical Engineering Department
Science University, University Town, UT 84123
Telephone: (999) 999-9999, E-mail: johng@scienceu.edu
URL: http://www.ee.science.edu/~johng
Education:
B.Sc., Math, Broad College, MI 48113, 1990.
M.Sc., EE, Maxwell University, New York, NY 12276, 1992.
Ph.D., EE, Maxwell University, New York, NY 12276, 1995. (**Advisor:** Nancy Albright)

Employment:
-Fall 2005–Present: Professor
-Spring 2000–Spring 2005: Associate Professor
Electrical Engineering Department
Science University
University Town, UT 84123
-Fall 2002–Fall 2004: Program Director
Division of Engineering

National Science Foundation
Arlington, VA 22230
-Fall 1995–Spring 2002: Assistant Professor
Electrical Engineering Department
Science University
University Town, UT 84123
-Fall 1990–Spring 1992: Teaching Assistant
Electrical Engineering Department
Maxwell University
New York, NY 10276

Publications related to the proposed research:
[1] J. Green. A new class of convolutional codes. *IEEE Trans. Inform. Theory*, pp. 14000–14010, June 2009.
[2] D. Joshi and J. Green. A novel data compression technique. *Int. Image Proc. Conf.*, pp. 16023–16029, Austin, Texas, August 2007.
[3] M. Lee and J. Green. Randomized transforms. *IEEE Trans. Image Proc.*, pp. 15672–15681, Feb. 2005.

[4] J. Green. A survey of discrete transform algorithms. *J. Signal Proc.*, pp. 14340–14350, May 2004.
[5] J. Green and M. Yuksel. Adaptive routing in wireless networks. *J. Wireless Commun.*, pp. 12345–12352, September 2004.

Five other publications:
[6] W. Fang, G. Brandburg, and J. Green. Rotational symmetries in discrete transforms. *J. Math. Models*, pp. 1022–1034. May 2003.
[7] J. Mukherjee and J. Green. Adaptive signal processing algorithms. *Trans. Signal Proc.*, pp. 9423–9433, April 2002.
[8] J. Green. Statistical techniques in image processing and pattern recognition. *Proc. of IEEE*, pp. 16720-16740, January 2001.
[9] J. Green and A. Denzelos. Randomized guessing and communication. *Trans. Commun.*, pp. 17345–17356, 2000.
[10] J. Green and A. Vincetelli. Fast discrete signal processing transforms. *Discrete Math*, pp. 13456–13466, June 2000.

Synergistic activities and professional service:
[1] Program Director, Engineering Division. NSF, 2002–2004.
[2] Associate Editor, *J. Signal Processing*, 2004–2008.
[3] Associate Editor, *Discrete Mathematics*, 2007–Present.
[4] Advisor for 20 doctoral and 32 master's students at Science U.
[5] Author of more than 55 journal articles and 105 conference papers.
[6] Inventor or coinventor in 11 patents on signal and image processing algorithms.

Invited talks:
-Keynote Speaker. Statistical Signal Processing. Int. Conf. on Signal Processing. Rome, Italy. August 2010.
-Invited Speaker. A new class of image processing algorithms. Workshop on Recent Developments in Image Processing, Boston, MA, June 2009.
-Keynote Speaker. Quantum signal processing algorithms. Int. Conf. on Quantum communications. San Diego, CA, September 2008.
-Invited Speaker. Generalized transforms and their applications in image processing. Workshop on Recent Developments in Image Processing, Boston, MA, June 2007.

Recognition and awards:
[1] *Who's Who in Science and Engineering*, 2008.
[2] ACM Fellow, 2006.
[3] IEEE Fellow, 2005.
[4] Putnam Fellow, 1987, 1988.

3.10   Ten most common mistakes
that make proposals fail

Many factors cause proposals to fail, but reviewers look for obvious imperfections to mark down a proposal. The following ten mistakes are quite common and must be avoided at all cost.

3.10.1   Mistake 1: Submitting a poorly
composed proposal summary

The summary of a research proposal is where the first mental contact is made between reviewers and the investigator who wrote it. It is where reviewers form their first and lasting impressions about a proposal. Therefore, it is where investigators have the least margin of error in making their cases. An erroneous claim about the intellectual merit or broader impact of the proposed work, inadequate treatment of any of the key components in the proposal summary, or failing to relate them logically would project an unfavorable picture of the proposal and likely cause its downfall.

3.10.2   Mistake 2: Avoiding preliminary results

It is surely efficient to get a research proposal funded with as little to show for it but many researchers set themselves up for a failure by being overly optimistic in this regard. It is in no one's best interest, including investigators', to waste time on a research problem if it is not likely to lead to significant new results. After all, wherever they may come from, funds are limited and should be used to support the most deserving research proposals. If all researchers act responsibly and use some of their own time to increase the probability of obtaining significant results, funds can be utilized more effectively. This will make the field more productive and efficient as a whole.

3.10.3   Mistake 3: Omitting a work plan

A sizeable number of investigators seem to underestimate the significant role that a work plan plays in the evaluation and awarding of a research proposal. Some investigators completely leave out the work plan and some view it as a short list of expected contributions. Yet, without a clear plan and schedule of deliverables, it is difficult to assess the tangible outcomes or contributions of a proposal, and this surely is enough to disappoint most reviewers and mark down the proposal.

3.10.4   Mistake 4: Submitting an inappropriate budget

Investigators must not let the budget get out of control or out of phase with the goals of their proposal and work plans. The dynamics of figuring an appropriate budget for the work packages described in a proposal can sometimes overwhelm investigators when coupled with overhead costs and various other budget line items. Iterative budget computations may force them to make changes in the budget figures without readjusting the work packages, leading to inconsistencies between the budget figures and proposed work packages. For example, if a work package requires a half-time research assistant, which is later removed to reduce the budget, then that work package must either be removed or modified. Any noticeable mismatch between budget figures and work packages will raise a flag and make the reviewers weary of the investigator's competence or intentions. This happens quite frequently and investigators must have an effective strategy to deal with it.

It is also important to remain within the bounds of announced budgets when submitting proposals. Ultimately, program directors must deal with their own budget restrictions, and it is easier to reject high budget proposals when all else fails.

3.10.5   Mistake 5: Submitting a poorly written proposal

Reviewers are not asked to evaluate proposals to determine if they are grammatically appealing or free of spelling errors, but such deficiencies often prevent them from making favorable evaluations. Citation brackets with incorrect, missing, or question mark entries, or erroneous cross-references to sections and formal statements can have a devastating effect on proposals and must be eliminated before the final submission of the proposal. Typically, these are caused by typesetting software used by investigators. Some typesetting software such as TeX and LaTeX require multiple runs to number citations and cross-references accurately in the document. Sometimes, the original cross-reference or citation codes may be miscopied. Multiple runs of TeX or LaTeX software cannot correct this type of error. The only solution is to read the proposal carefully to see if the cross-references and citations are semantically correct as well. These errors can also occur in other document typesetting systems. The Microsoft Office Word software is particularly notorious for changing or mixing up index values for sections, theorems, figures, and tables. Extreme care must be used to avoid such indexing errors.

Transitions between sentences, paragraphs, sections, and other bodies of proposals are just as critical. Careless connections or transitions between adjacent document bodies may easily annoy reviewers and affect their evaluations of proposals. A good way to avoid such problems is to read the proposal carefully after it is completed, and if possible, also ask a colleague to read it with a critical eye.

3.10.6 Mistake 6: Marginalizing or downplaying someone's work

There is no doubt that scientific progress is fueled by the work of competitive researchers that find the current state of scientific knowledge incomplete one way or another. Building on the contributions of others is essential for scientific progress. However, this does not mean that one should be indiscrete in characterizing the accomplishments of fellow researchers. It is helpful to see the glass half full rather than half empty, and give credit to the contributions of others while generating new ideas and reporting new results. Nowhere is this more critical than in the process of writing proposals, as being less than courteous to an earlier study can easily antagonize a reviewer who may be connected to that study or know someone who does.

The following statements are examples of marginalizing someone's work that may not appear to sound offensive when viewed subjectively:

The claims given in [35] cannot possibly be true because Smith's earlier work has flaws.

The statement given in [17] by Jackson does not hold when the temperature is too high.

The experiment that Anderson described in [23] does not make sense since it was not verified.

These statements are not only offensive but they are also imprecise and unscientific. For example, in the first statement, how is a reader supposed to know that Smith's earlier work has flaws? In the second statement, what value should the reader assume about the temperature mentioned? In the last sentence, how does the author know that Anderson did not verify the results of his experiment?

Obviously, there should be room for criticism of earlier work, but this should be done scientifically and without using derogatory statements. Each of the aforementioned flaming statements can be revised to make it a robust scientific remark and eliminate its offensive tone. For example, the first statement can be modified as follows: "Results given in [35] may work under the assumptions given there, but we will not use these assumptions here." Similarly, the last statement can be reworded as follows: "It is difficult to replicate the experiment described in [23] without knowing the exact conditions under which it was carried." Thus, a good rule of thumb is to avoid using names in citations, where the connotation of the reference is not favorable to the author of the cited source.

3.10.7   Mistake 7: Not acknowledging related work or plagiarizing it

In this day and age of the World Wide Web and Internet, it is embarrassing and nearly inexcusable to play innocent when omitting any work that is relevant to a research proposal; it is downright wrong and unlawful to plagiarize it. In order to minimize the risk of omitting a relevant work unintentionally, one should perform an extensive literature survey using search engines and bibliographic databases.

In some fields, proposals may be evaluated by reviewers who choose to present and publish their work in particular conferences and journals. This may create a review atmosphere that may be skewed favorably toward proposals submitted by investigators who present and publish their work in the same conferences and journals. Although this may not be fair, it is often unavoidable. In this case, a good strategy would be to make sure that the significant contributions are cited in the proposal regardless of their authors, the conferences they attend, or journals in which they publish their work. Of course, one needs to be discreet in incorporating references to a research proposal and not cite a source just to pamper a potential reviewer unless it is relevant to the proposed work.

3.10.8   Mistake 8: Citing dated references

There is a general sentiment among reviewers that a proposal that relies on old references cannot produce new results, and it is likely to recycle known results unknown to its investigator. There may be exceptions to this observation, but as noted in Section 2.5, research proposals must establish some affinity to the field in which they seek to generate new ideas. A good indication of demonstrating this affinity to existing research is to cite references that are recent and relevant to the proposal's research field. Besides being important to demonstrate that the proposal is up-to-date with its references, it helps to actually read the contemporary articles to strengthen the need for carrying out a proposed research.

It is not very likely for two investigators to have the exact same point of view or set of goals even though some degree of overlap between research projects is always present. This is precisely where the art of writing good proposals makes all the difference by exposing the ideas in a different light and making them distinct from previously studied problems. Of course, this is difficult to accomplish unless a researcher is up-to-date with the most recent results in the field. Conference papers play a more critical role in keeping up with recent results, as the backlogs of journals keep many researchers from following the key results published in journal articles. This is changing slowly with emerging online publication sites such as www.arxiv.org that provide rapid publication capabilities for researchers to fully disseminate their ideas as quickly as possible.

3.10.9   Mistake 9: Downplaying the impact of investigators' credentials

A common thread of discussions on proposal review panels deals with the qualifications of investigators as their proposals come up for evaluation. Any time a proposal is submitted by a researcher who is not an established name in the field, the discussion gets more heated, and in the end, the outcome is rarely in favor of the researcher in question. This is the main reason for funding agencies to insist on some track record in their solicitations. Even submitting an outstanding proposal may not persuade some reviewers if an investigator is not established as a credible researcher in the field. Junior investigators with fresh Ph.D.s may be the only exception to this rule.

The best remedy to the credibility issue is not to submit a proposal until a credible level of recognition is accomplished. Of course, judging one's credentials to conduct research in a field is not easy. Optimism takes over when researchers judge their own qualifications, and when they judge others, pessimism takes over. The truth must be somewhere in between and researchers must be as critical of their own qualifications as those of fellow researchers. In Chapter 2, we provided some guidelines for determining one's qualifications in a field, and these guidelines should be used to arrive at a realistic evaluation.

Reviewers often grant some leeway for researchers with a strong track record in adjoining research areas and may overlook the lack of productivity in the immediate area of the proposed research. In fact, this is one of the reasons funding agencies like NSF ask investigators to list up to five additional publications in any field of research. Clearly, such publications, to the extent they stray from the subject of proposed research, provide only a baseline of credibility. As suggested in Chapter 2, a good strategy is to hold off submitting a proposal until publishing at least one full article in an archival journal and presenting preliminary results at a major conference or meeting, both related to the subject of the proposal. Collaborative proposals are another way to minimize the potential downfall from a lack of qualifications but care must be used to make sure that the qualifications of the investigators are balanced and cover the key needs of proposed research.

3.10.10   Mistake 10: Submitting a work plan with infeasible resources

Resources, resources, and resources! Carrying out research all boils down to having the right resources with new ideas and a matching work plan.

Funds are there to trigger the proposal into action but they are not sufficient to magically create the scientifically competent human resources and critical equipment and tools. These have to be available when and if the proposal is funded to make it feasible.

The entire team of researchers including postdoctoral scientists, graduate and undergraduate assistants, programmers, and technical support personnel all factor into the final decision when it comes to funding a research proposal. Investigators must make sure that they have the best and brightest minds that can be found in their departments and institutions. For highly ranked institutions, this may not be a serious problem, as their standings in national rankings would attract the highest caliber of graduate students and postdoctoral scientists. In middle and low ranked institutions, constituting such a high quality research team and demonstrating the quality and talent of human resources may be more of a challenge. Here again, this is where working and collaborating with graduate students and postdoctoral researchers in a preliminary research study prior to submitting a proposal becomes critical. Mentioning that some number of graduate students will be hired to work on the proposed research alone is often not good enough. Their qualifications as they relate to the proposed research can play a significant role in the funding of a proposal.

The equipment and tools requested in a proposal are also critical to winning or losing funding, and their availability for use in the proposed research must be demonstrated. For example, an electron microscope located in a laboratory in some department on campus may not be as accessible as the one in an investigator's own laboratory and reviewers must be persuaded that potential scheduling and access conflicts for utilizing equipment and tools will not hamper the proposed research. The technical resolution of the proposed equipment is another contentious issue that causes debates between reviewers if the resolution can handle the measurements and experiments to be conducted in the proposed research. If any assumptions or provisions are made for overcoming the limitations of the available equipment and tools to carry out the proposed measurements and experiments, they should be stated explicitly, and the reviewers must be persuaded that such assumptions remain within the margin of error of the expected measurements and results.

3.11   *What to do if your proposal is declined*

Statistically, odds are not in anyone's favor to receive funding, since funding rates by most agencies rarely exceed 30%, and 20% to 25% are more typical. Some programs award proposals at rates even below 20%. Hence, it is more than likely that almost all investigators who routinely submit proposals to receive funding have had one or more of their proposals declined, and this author is one of them. This is the "do not feel bad because it happens to most of us" side of being notified for a declination of a proposal. The not-so-pleasant side is that a declined proposal means that the funds will not be forthcoming and the research you had planned to conduct may have to be delayed without the funds, and worse, it may have to be dropped altogether.

Most investigators experience some degree of disappointment when their proposals are declined. Some may take it personally and blame the system for not being fair or objective. Others may quietly shy away from the research field and switch their research field or even go do something else and leave research behind them altogether. Still others may look for legal venues to prove that their work was badly judged or unfairly handled. All these are normal reactions within reasonable bounds and can help researchers ease their frustrations. In many ways, they are no different than how most people react when they feel that they are unfairly judged or treated. After all, no credible investigator will knowingly submit a bad proposal and has every expectation that fellow researchers in the field will appreciate the good ideas laid out in his or her proposals. However, the reality is that there will almost always be more proposals in a pool than the monetary resources available to fund them.

So, what should you do if your proposal is declined? The following are some of the steps that you may wish to include in your short list of actions:

Go over the reviews carefully and check if there is a consensus among the reviewers about the apparent weaknesses of your proposal. Momentarily, switch your role from that of an investigator to that of a reviewer and see if you would agree with any of the criticisms of the reviewers. If you do, make a list of changes that you would do to avoid these criticisms. Otherwise, make a list of your objections by staying away from personal remarks as much as possible.

Break the news to your postdoctoral colleagues and graduate students, and share the reviews with them. Acknowledge that this is a setback but also assure them that there will be other funding opportunities and they should remain optimistic as you do. You can even let them know that funding statistics normally work against all proposals, but some proposals in every proposal pool must actually make those statistics.

Contact the program director that handled your proposal in the funding agency and arrange a meeting to discuss your proposal. The goal of this meeting ought to be to share your thoughts with the program director and let him know if you agree with the criticisms of reviewers as well as state your objections. Maintaining a scientific atmosphere of exchange while expressing your opinions will be helpful in leaving a positive impression with the program director. That is the only tangible goal that your meeting can potentially accomplish at this point, and any feedback from the program director would be a bonus for preparing the next proposal.

Some program directors may extend invitations to investigators to review proposals or serve on review panels. If such an invitation does not prevent you from submitting a proposal during the next cycle, it will be a good idea to accept the invitation. In general, it is helpful to serve on review panels to see how fellow researchers judge proposals. There is no better substitute for seeing how reviewers reach and tender their recommendations.

Once the initial period of reactions and responses is over, it will be time to decide whether to revive the declined proposal and submit it for funding again. Funding agencies often discourage investigators from resubmitting proposals without a substantial revision. Some may also require an explanation of the changes when such a resubmission is made. Overall, it is safe to assume that a substantial revision is expected before most funding agencies would entertain another submission.

The revision of a declined proposal must clearly address all the issues raised by the reviewers and introduce compelling fixes that cannot easily be refuted or challenged. The process of revision is not as structured as in revising a manuscript submitted to a proposal, but there are similarities. Common criticisms such as incomplete or inappropriate references can easily be fixed. The weaknesses in problem statement and approach parts can also be fixed relatively easily if the criticisms are mild. If they challenge the originality of the research problems or the approach then more effort will likely be required to strengthen the proposal.

If the criticisms of reviewers are directed at the qualifications of the investigators, even a substantial revision of the proposal may prove ineffective, and you may wish to consider postponing a resubmission until you generate some preliminary results and gain some traction with other researchers in the field. The review process generally has a memory of three to four review cycles, and this may provide a sufficient time to beef up your publication record and also further develop the ideas in your original proposal before submitting it again.

After all these considerations, if you still feel that your proposal has been unfairly treated by the reviewers, a prudent course of action might be to appeal the decision even though the available statistics on appeals, at least in the case of NSF, are not in favor of obtaining a reversal through such a process, as will be discussed in Chapter 6.

3.12   Summary

This chapter presented the key building blocks of a research proposal, and explained how to build a proposal in step-by-step fashion from the summary to expected contributions using the proposed work, intellectual merit, and broader impact as the anchors of the proposal. Also presented were the 10 most common mistakes that might cause proposals to be declined and suggested what can be done to revive the ideas in a proposal in case it is declined.

An important conclusion that can be drawn from this chapter is that although there is no unique approach to writing a winning proposal, such a proposal is made up of essential parts such as problem statement, intellectual merit, broader work, approach, work plan, and investigators' qualifications that must all be skillfully fused together to withstand the criticisms of most stringent and rigorous reviewers.

3.13   Bibliographical notes

The awards summaries that have been discussed in the chapter have been obtained from NSF's Awards site.* This site provides access to the summaries of all NSF awards and should be useful for any investigator to keep track of what NSF funds and what other investigators work on in any of the fields that are funded by NSF.

3.14   Questions

3.1   How do you begin writing your proposal? Which section is most difficult for you to write?

3.2   How closely does the proposal outline in Section 3.1 match your own structuring of a proposal? How will you modify it to make it more effective for presenting your ideas in a proposal?

3.3   Formalize your ideas in your field of research into a skeleton of a research proposal and test its intellectual merit, using the flowchart in Figure 2.1.

3.4   Do you use one of the proposal summary styles described in the chapter in your proposals? Which style matches your research proposals best?

3.5   Would you submit a research proposal in a new area of research without first publishing your preliminary findings?

3.6   If a researcher has just published a result that overlaps with your ideas in a proposal you are about to submit, how will you handle it in your literature survey?

---

* NSF Award Search, http://www.nsf.gov/awardsearch/.

3.7 When you write a research proposal, do you mostly recycle your references from your earlier proposals? How do you keep up with more recent research results?

3.8 If you were faced with a budget constraint that prevented you from conducting your proposed research adequately, what would you do?

3.9 Do you see any potential downside or risk in citing too many references in a research proposal to make sure that everyone's work is credited?

3.10 If you think that not having good qualifications prevents you from submitting a research proposal, state three things you should do in the short, middle, and long term to help alleviate your situation.

3.11 Do you think that a work plan plays a role in the success of a research proposal? Do you include a work plan when you submit a research proposal?

3.12 Is there any relation between expected contributions and broader impacts of a research proposal? Are there any differences between them?

3.13 Sometimes reviewers will question whether a proposed research is feasible. How would you address this question in a research proposal? What sections of the proposal will have to be revised to address it?

3.14 Can you think of any mistakes other than those listed in Section 3.9 that may cause a proposal to fail?

3.15 If your proposal is declined, what will you do? Will the suggestions described in the chapter work for you? Are there other things you can do to increase your chances of funding next time?

chapter four

Getting on with conducting a funded research project

One of the most positive aspects of winning an award is the feeling of optimism investigators get about the versatility of their ideas. This can help propel a research project, but it also carries the risk of quickly degenerating into a self-aggrandizing attitude to lose sight of the work that lies ahead. In this chapter, the aim is (a) to discuss the generic tasks that are commonly carried out while conducting funded research in universities, and (b) to describe some of the key activities that may affect the success of a research project, including how to transform a proposed research into a work plan, design work schedules, hire research personnel, and streamline the publication of research results.

4.1   Assembling a research team

Once a research proposal is officially awarded and funds are transferred from the funding agency to a spending account at the home institution of an investigator, the research team can be assembled to launch the project. In some research proposals, research teams are already identified down to postdoctoral researchers and graduate research assistants. In other proposals, only the types and number of positions in a research team may be specified. In either case, funding agencies often ask investigators to revise their budgets downward. This will likely cause investigators to rethink their hiring priorities and reassemble their research teams with available resources to accomplish the goals of their research projects.

4.1.1   Graduate students

Funded research projects in a university setting attract graduate students like a magnet. They provide them premium research assistantships, prestige, and almost a sure path to a career involving research in one form or another. Still, investigators often compete to draw talented graduate students to their research programs since there is a limited pool of such students on any campus. Recruiting research assistants is further compounded by the fact that top students would like to optimize their choices. This may frustrate investigators who are just looking to get their research underway.

Nonetheless, recruiting research assistants should be as rigorous and competitive as possible. Investigators can publicize the available research positions with a succinct description of the subject of their research projects and the nature of the work that will be performed by graduate research assistants. This will signal all potential candidates, including the talented ones, that they are competing for limited resources, and not having the privilege to work on a research project is not a good prospect. In a way, this is analogous to the proposal submission process where investigators must put forth their best effort to compete with fellow investigators.

Another venue for recruiting graduate students is to teach advanced graduate courses. Most departments promote and offer such courses. Investigators should introduce new topics courses as they move their research frontiers. Even if offering such courses is not an option, teaching graduate courses provides an opportunity to meet new graduate students and find about their research interests. A good rule of thumb for active investigators is to teach one graduate course per year to remain in touch with the evolving body of graduate students in their departments.

When recruiting new graduate students to a new research team, an important point to keep in mind is that there should be a good match between the research interests and professional plans of the potential research assistants and the goals of the research project. Hiring students merely because of their superior performance in course work and without probing their interests in a research project may ultimately waste the precious resources of the project and impact its success. A better alternative would be to interview a small pool of qualified students and offer them research assistantships based on the following criteria:

- Competence to handle the work
- Interest in the work
- Professional goals

Ideally, candidates should rank highly competitive with respect to all three criteria but they can be weighted to match the priorities of a research project. Here, there are potential trade-offs. Some investigators prefer to hire students who are able to deal with building complex systems and experiments, and conduct original research at the same time. Students who have been involved in system building or oriented toward hands-on projects will likely be more effective in such research projects. Other investigators may be in need of theoretically motivated students to carry out their projects. Again, the interpretation of the phrase "competence to handle the work" must be coupled with the type of research and the nature of work to be performed. Of course, a research project requires a combination of different types of skills and experiences. Investigators should carefully adjust their hiring decisions to create the right balance of skills needed.

Timing plays a critical role in forming a formidable research team in a university setting. Often, students apply for admission to graduate programs late fall each year. The applicants with the best credentials are identified by graduate studies offices in many departments and presented to the faculty for early fellowship offers with matching funds from research projects. This provides a good opportunity for investigators to identify the qualified applicants that they are interested in hiring as graduate research assistants. They can conduct e-mail and phone interviews with such prospective students before committing their resources. Typically, applicants do not decide on their offers until late spring and this gives investigators ample time to continue to interact with promising applicants. This approach should work well for investigators with new research funding or uncommitted funds.

In those cases, where funding arrives later in the year than the spring, investigators have two choices: They may find a few potential candidates among the current graduate students without financial aid. Not all of these students may be as well-qualified academically, but some may be competent to handle the tasks needed to be carried out in a research project and have a keen interest in the research project itself. Others may have particular weaknesses in their educational preparation due to curriculum differences and for other reasons, and this may limit their impact on the project if hired. To increase the likelihood of finding the most qualified candidates among this pool of students, a more thorough interview and assessment process can be used. The second alternative would be to wait for the next round of admissions and hire from the pool of top applicants for the next fall. Depending on the duration and work plan of the research project, this may or may not be feasible, and a combination of the two alternatives may have to be used.

4.1.2   Undergraduate students

Undergraduate students are among the most underrecruited human resources in university research projects. Most research-oriented universities make an effort to establish undergraduate research programs to motivate their research faculty to work with undergraduate students. Some funding agencies provide additional funds, such as the Research Experience for Undergraduates (REU) program at the National Science Foundation (NSF), to support such programs as well. However, there are some potential risks in hiring undergraduate students that may make some faculty members cautious. With such undergraduate research programs, an impression is created that everyone should be involved in research projects, and some faculty may feel this to be an undue pressure on them.

Furthermore, investigators may find their workload to increase for training such students without having any impact on their projects. In addition, undergraduate students have tight class schedules that leave very
little time for them to focus on a research project that requires creative
thinking and persistent effort. When a winter or summer break comes,
undergraduate students quickly disappear and any projects to which they
are assigned will have to wait until they are back from their breaks.

Despite these potential drawbacks, undergraduate students may fare
better than graduate students for some projects, both in talent and cost
terms. They may do especially well in research projects that involve well-
defined tasks such as programming projects, performing and monitoring
experiments, and carrying out field studies. Occasionally, investigators
may also encounter gifted undergraduate students with a strong interest in conducting original scientific work. Identifying such students may
seem difficult at first, but an announcement or two in undergraduate
classes or relevant student societies will bring them out and connect with
investigators.

4.1.3   *Postdoctoral researchers and research scientists*

Some research projects include postdoctoral and research scientist positions in their budgets. One clear advantage of hiring a postdoctoral
researcher in a project is the experience and professional preparation that
comes with the completion of a doctoral degree. Postdoctoral researchers
are already trained to conduct research and no time would be lost to get
them started on the project. Graduate students, on the other hand, will
need time not only to learn about the state-of-the-art research results in
the field but also to develop skills to formulate and solve research problems. Of course, the cost of hiring a postdoctoral researcher will be higher
than for a research assistant even though, with increasing competition
between universities, graduate student fellowship packages are becoming
competitive with postdoctoral salaries.

Postdoctoral researchers can come both from within and outside
an investigator's home institution. Some postdoctoral researchers in the
home institution may be flexible to work on more than one research project, and their availability should be explored to match the needs of an
available project. To increase the effectiveness of a postdoctoral research
position, an investigator may have to look at postdoctoral researchers at
other institutions as well. Such candidates may be found among the pool
of graduate students who are nearing their doctoral programs. They may
also be working as postdoctoral researchers and approaching the end of
their contracts.

Investigators often reassign their postdoctoral researchers and other
research personnel to a new research project when they receive new

funding. Such reassignments are done routinely in large research projects managed by research institutes and centers to keep their research personnel employed. There is a certain degree of risk in this approach as it may generate a climate where people come to expect that they can switch to other projects so as not to lose their positions. Investigators who work in research centers and institutes should be especially careful to protect their research projects from aggressive managers and administrators when they secure their own funding to carry them out. In some cases, a new research project for which an investigator receives funding independently may require the resources such as laboratories and cleanrooms provided by a research center. In these cases, the budget would ordinarily include support for personnel expenses such as technicians and engineers. Beyond these, hiring of any new person to a research team should be done solely on the needs of the project, and investigators should be given the freedom to hire their research personnel. Administrators should not forget that the success of a proposed research hinges on the research team and a successful completion of a research project serves everyone's interest in their institutions.

4.2   Carrying out the proposed work

Research is clearly an open-ended process. However, it is not a random walk and needs a good plan to reach its milestones. There are certain commonsense practices that can be useful in planning and conducting research projects as described next.

4.2.1   Developing a work plan

It was mentioned in Chapter 3 that an effective work plan is critical to the success of a proposal. Once a proposal is funded, the proposed work plan can be further refined and adjusted to factor any revisions that were made prior to signing a contract with a funding agency. The sample work plan that was described in Section 3.7 applies to projects that can be divided into tasks with predetermined completion times. Typical examples of such tasks include laboratory experiments, design and simulation projects, and field studies. The project time is divided into equal time periods during which some of these tasks can be performed concurrently while others can only be executed in tandem. The primary goal of drawing up a work plan is to match the human and material resources and the duration of the project as effectively as possible. Here the effectiveness implies using the resources in such a way so as to accomplish the goals of the project as expeditiously as possible.

In a more general setting, a work plan may be pictured by a 3-D diagram as shown in Figure 4.1. The three axes represent the "team members,"

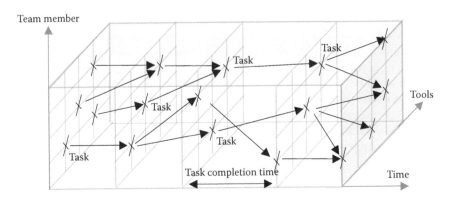

Figure 4.1 A 3-D view of a work plan.

"tools," and project time. The tools can be viewed as any equipment or facility extending from handpicks, dustpans, shovels, trowels, scoops, sifters, bulldozers, and mapping equipment used by archaeologists and anthropologists, to desktop and laptop computers, electron microscopes, radiological probes and scanners, magnetometers, telescopes, experimental benches, test beds, cleanrooms, particle accelerators, and colliders that are used in research projects in physical sciences. They can even be simple pencil and paper, as researchers in theoretical fields would hardly need anything else. Some of these tools may be available freely at any time but using others may be constrained by their availability since they may have to be shared or operated with certain restrictions.

The points in the diagram represent the tasks to be performed by the members of the research team using the tools and facilities to which they can be assigned without any conflicts. Some tools, such as shovels, scoops, and microscopes, can be used by one researcher at a time. Other tools, such as computer servers, test beds, and cleanrooms, can be shared by several researchers. It is also possible for a researcher to utilize several tools at once. These cases correspond to assignment patterns as seen in the leftmost plane in the diagram. Some researchers in the team may work without any tools and, likewise, some tools may be operated without any human intervention for a period of time. If no tools are used in a research project, then the work plan diagram reduces to a plane. If all researchers can access the tools without any conflict any time, the work plan can again be simplified to a two-dimensional representation.

The project period is divided into time slices and at the beginning of each time slice, team researchers may be assigned new tasks and matched with the tools or they may continue to work on a project they had begun earlier. The initiations of the time slices are represented by the grid planes, which are perpendicular to the time axis. The points on the grid planes represent pairings of team members and tools. Ideally, all researchers must be kept busy and all tools must be in use within their operational capabilities and subject to potential dependences between the tasks. The arrows indicate the dependences between the initiations of tasks on different grid planes. A task on a given grid plane cannot be initiated until all the tasks on the prior grid planes with arrows pointing at it are completed. The durations of time between consecutive grid planes specify the task completion times, that is, how long the tasks initiated on the originating grid planes take to complete. The sum of all the times between consecutive grid planes gives the project completion time.

Tasks in a project may be subject to certain constraints with regard to the resources available for the project. For example, a research project may have all the tools it needs but it may have a small team, or it may have limited tools with a large team of researchers. Both cases could result in a bottleneck of resources and may push the project behind its intended schedule. A competent investigator should be able to match the capacity of his or her research team with the available tools and equipment.

In general, designing a work plan involves deriving a schedule by which the proposed research can be completed with the available resources by a desired date or within a desired period of time. This design process can manifest itself in several ways and with several degrees of freedom that must be traded against one another. The dependences between the tasks in a research project present one such degree of freedom. These dependences are often formed during the partition of the proposed work into smaller work packages or tasks, and can be revised during iterations of the work plan. The current research teams of principal investigators and others that may be hired during the project provide another degree of freedom. The tools in their laboratories or other facilities they can access during the project constitute still another degree of freedom. Finally, the duration of the project is also a degree of freedom that they can possibly revise and trade against other degrees of freedom. It is reasonable to assume that most researchers would like to accomplish their goals as quickly as possible, but, besides possible scientific setbacks, the potential constraints on all these degrees of freedom, including budget restrictions, may set limits on the completion time of research projects.

4.2.2   Using resources effectively

So, how do principal investigators deal with all these complex issues related to designing an effective work plan that allows them to accomplish the goals of their research projects? Initially, the whole project is one major task, W, and the investigator of the project has agreed to accomplish this task within a set period time, T. Thus, the project must be completed at the rate of $X = W/T$ tasks per predetermined unit time in order to be completed in the time budgeted in the proposal. In order for this to happen, the investigator will have to find a way to divide the scientific challenges stated in the proposed research into a number of tasks that can be performed within predetermined time slices without going over the budgeted time. For example, if the initial task is divided into 12 sub-tasks, each of which takes 1 month to complete, and only one task can be performed during each month, the entire project cannot be completed in less than a year. Stated another way, it is not feasible to complete such a research project at the rate of X tasks per month, for any $X < 1$.

Thus, a reasonable first step would be to begin to identify the key tasks such as creating mathematical models to capture the crux of research problems, formulating analytical solutions, setting up experiments, and designing simulation models, and to tie these tasks together logically to accomplish the goals of the proposed research without being overly con-cerned about the dependences between them. If a work plan was included in the proposed research, it can provide a first pass at this divide–conquer process.

One way or another, this initial partitioning of the proposed research should result in what will be referred to as a *task flow graph*. Task flow graphs are directed and acyclic (cycle free) interconnections of tasks where the directed arcs indicate the dependences between the tasks as shown in Figure 4.2. The task flow graph on the left describes a project in which the tasks are designed to have maximum dependence between one another. The project begins with task W_1 and ends with task W_6, and all tasks in between are sequentially completed. In this case, it does not matter how many researchers are assigned to the project, and the duration of the proj-ect is given by the sum of the processing times of all the tasks. Thus, if each task takes 2 months to complete, the project cannot be completed in less than 12 months whatever sources are used. This type of task graph describes a research project that involves a single investigator without any other research personnel. When the research team includes more than one

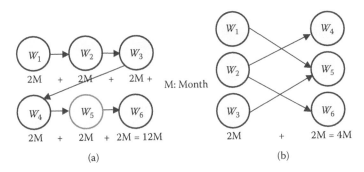

Figure 4.2 Examples of task flow graphs.

researcher, the same task graph inevitably leads to an inefficient partition of the proposed research, as it will make some researchers remain idle unless the same six tasks are repeated several times. Although this may work for some task partitions that exploit pipelining, such a technique can hardly be applied to handling tasks in research proposals.

The second task flow graph in Figure 4.2b is designed to allow some degree of independence between the tasks. More specifically, each of the three tasks on the left can be initiated and completed independently of the other two tasks, and likewise, each of the three tasks on the right can be processed independently of the other two tasks. However, the three tasks on the right cannot be initiated until one or more of the tasks on the left are completed. These conditions permit the project to be expedited if the research team consists of at least three researchers.* In particular, suppose that each task takes 2 months to complete as before. Then, with three researchers working concurrently, the project can be completed in 4 months rather than 12 months as compared to the task flow in Figure 4.2a.

Once a preliminary task flow graph is designed, the next step would be to map this graph into a *work flow graph* where the tasks are matched with the available resources within a time frame. The difference between a task flow graph and a work flow graph is that the former only describes the temporal dependences between tasks, whereas the latter describes both temporal dependences and resource constraints (dependences). For example, in the task flow graph in Figure 4.3a, the temporal dependences allow the project to be completed in four months but this requires three researchers. If we use only two researchers, we cannot take advantage of the fact that up to three tasks can be performed concurrently and effectively, the resource constraint forces the tasks to be performed less concurrently, leading to a project completion time of six months as shown in Figure 4.3b.

The thicker arrows in both graphs in Figure 4.3 indicate the dependences between the same six tasks in a possible corresponding work flow graph assuming that only three and two researchers are assigned to carry out the tasks, respectively. These resources can be the investigator, a research assistant, or pairing of a research assistant with a tool or workbench. The numbers in elliptic regions indicate the order of completion of tasks in the work flow graphs. Any task flow graph can be transformed to one or more work flow graphs where temporal dependences remain enforced and resource dependences are satisfied. This can be seen in both work flow graphs in Figure 4.3, where all the temporal dependences between the clusters of tasks generated in the transformation from the task flow graph to the work flow graphs are enforced.

---

* Here we assume that each task requires a separate researcher.

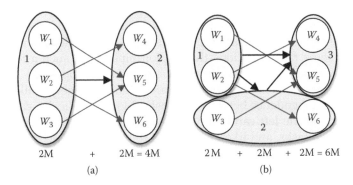

$$2M \quad + \quad 2M = 4M \qquad\qquad 2M \quad + \quad 2M \quad + \quad 2M = 6M$$

(a)   (b)

Figure 4.3 Transforming a task flow graph into work flow graphs.

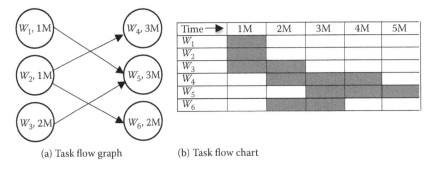

(a) Task flow graph   (b) Task flow chart

Figure 4.4 Weighted task flow graph and chart.

Task flow graphs such as those in Figure 4.2 can be helpful in visualizing the potential bottlenecks in reducing the completion times of projects, and redesigning and revising the tasks if needed. Ideally, the project must be divided into tasks with as few dependences as possible so that they can be initiated without having to wait for other tasks to be completed. Tasks should also be designed for equal completion times as much as possible. When this is not possible, the task graphs can be labeled with completion times of individual tasks as shown in Figure 4.4. The task flow chart provides a more explicit representation of the temporal dependences between tasks in a research project even though the two representations contain the same information. For example, the chart on the right immediately reveals that the temporal dependence between task W_3 and W_5 force the completion time to five months. The same information can be gathered from the task flow graph on the left by finding the dependence with the longest time period. With these two models, it should be possible to minimize or considerably reduce the completion times of projects and generate effective work plans.

4.2.3   Example

Consider a research project that involves conducting some 14 experiments, each taking 2 months from set up to completion. Furthermore, the expected results of these experiments will be presented in four meetings that will take place within 24 months after the project begins and the preparation of each of these presentations also takes 2 months. If the project can take no more than 12 months to complete and 3 research assistants are to be employed to conduct the 14 experiments and help prepare the papers to report the results, design a work plan so that all tasks (experiments and publications) planned in the research can be completed in 12 months.

Solution: It should be first noted that:

1. Since there are 18 tasks altogether each taking 2 months, 36 months are needed to complete all the tasks.
2. With three research assistants, each working for 12 months, there are 36 months of work force.

Thus, there are numerically enough research assistant months to complete the project in 12 months. Whether a work flow can actually be designed to accomplish this depends on the dependences between the tasks in the task flow. If the tasks are completely independent, then we can use the work flow shown in Table 4.1 to complete the project in 12 months where the E_i's and P_i's denote the experiments and presentations, respectively. This is the best we can do since we used all available time and manpower allocated to the project. This schedule will also work under certain work flow constraints. For example, if the 14 experiments can be partitioned into five groups, where (a) each of the first four groups consists of 3 experiments and the last one consists of 2 experiments and (b) the interdependences are limited to only those between the groups as shown in Figure 4.5, then a similar work flow schedule can be used to complete the project in 12 months.

There exist some task flows from which a work flow taking 12 months may not be constructed. The most obvious example is a task flow where

Table 4.1 Scheduling 14 Experiments and 4 Reporting Functions in 12 Months with 3 RAs

	M_1	M_2	M_3	M_4	M_5	M_6	M_7	M_8	M_9	M_{10}	M_{11}	M_{12}
RA_1	E_1	E_1	E_4	E_4	E_7	E_7	E_{10}	E_{10}	E_{13}	E_{13}	P_2	P_2
RA_2	E_2	E_2	E_5	E_5	E_8	E_8	E_{11}	E_{11}	E_{14}	E_{14}	P_3	P_3
RA_3	E_3	E_3	E_6	E_6	E_9	E_9	E_{12}	E_{12}	P_1	P_1	P_4	P_4

Note: RA_i: *i*th research assistant; M_i: *i*th month; E_i: *i*th experiment; P_i: *i*th presentation.

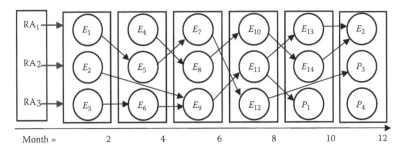

Figure 4.5 Clustering experiments into a 12-month work flow.

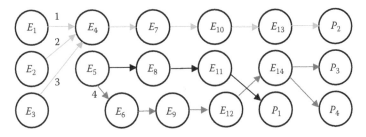

Figure 4.6 An infeasible task flow with three research assistants and 12-month completion time.

each of the 14 experiments is designed to depend on the completion of another experiment. This will take 28 months regardless of how many research assistants are used.

There are many other task flows for which a 12-month work flow is not feasible. Note that in the task flow diagram in Figure 4.5, no path consists of more than six tasks. Any path of more than six tasks in the task flow will obviously make a 12-month work flow infeasible. However, not all task flows in which all paths consist of six or fewer tasks result in 12-month work flows.

To see why, consider a task flow in which more than three paths of six tasks exist even though there is no path of more than six tasks. We wish to create a work flow with three research assistants as before. As an example, a task flow graph with a set of five paths of six tasks is shown in Figure 4.6. Since there are only three research assistants, one of E_1, E_2, E_3, or E_5 must be initiated after the other three tasks have been initiated. This means that either P_2 or P_3 and P_4 cannot be completed until the end of the 14th month assuming that each task takes 2 months as before. The first case occurs when one of E_1, E_2, E_3 is not assigned to a research assistant in the first round. The latter case occurs when the same three tasks are all assigned to the three research assistants in the first round, and the initiation of E_5 must wait for them to be completed.

4.2.4   Optimal work flow schedules using two researchers

Every task flow with u tasks, each taking v months can be transformed to a work flow that can take between uv/s and uv months using s resources. A work flow w performing a task flow using s resources is called optimal if it takes at most as long as any other work flow that performs the same task flow using s resources. In some specific cases, task flow graphs can be transformed into optimal work flows. In nearly all of these cases, all resources are assigned to tasks for the same length of time, and schedules that rely on this assumption are referred to as equal processing time schedules. In what follows, we consider two such schedules.

The first scheduling algorithm works for any task graph and two researchers with similar capabilities. The two resources may be two research assistants assigned to a project or two postdoctoral researchers. The algorithm was developed by E. G. Coffman and R. L. Graham for scheduling tasks on processors* and can be applied to transform any task flow graph in which all tasks are projected to take the same amount of time into a work flow graph with clusters of size one or two. It was shown that this algorithm produces an optimal work flow for any task flow if only two resources are used.

The algorithm uses a recursive labeling scheme that marks the nodes in a task flow graph using an ordering relation between the lists of labels of the successors of tasks. To facilitate the description of the algorithm, let us call a task W_j a successor of task W_i if there is an edge directed from W_i toward W_j, and call W_j a predecessor of task W_i if there is an edge directed from W_j toward W_i.

The Coffman–Graham algorithm is best explained by an example, and so we consider the task flow graph in Figure 4.7a. The tasks are originally labeled inside the circles from left to right and top to bottom in that order. Given that there are 13 tasks, and two researchers are allowed, the task flow cannot be completed in less than 14 months, assuming each task takes 2 months to complete. However, because of the dependences between the tasks, it will likely take longer to complete this task flow. Exactly how much longer is answered by Coffman–Graham algorithm.

---

* Coffman, E. G, and Graham, R. L. 1972. Optimal scheduling on two-processor systems. *Acta Informatica* 1:200–213.

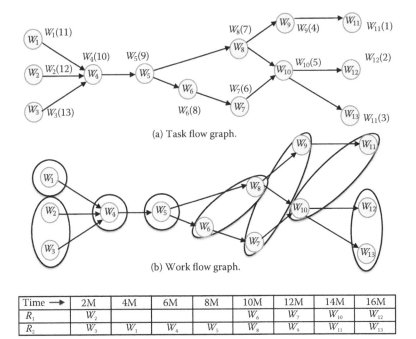

(a) Task flow graph.

(b) Work flow graph.

Time →	2M	4M	6M	8M	10M	12M	14M	16M
R_1	W_2				W_6	W_7	W_{10}	W_{12}
R_2	W_3	W_1	W_4	W_5	W_8	W_9	W_{11}	W_{13}

(c) Work flow schedule with each task taking 2 months to complete.

Figure 4.7 Task flow and corresponding work flow and schedule using two research assistants.

The Coffman–Graham algorithm first computes a permutation of the task labels using the following procedure. This labeling is then used to assign the tasks to the two researchers as will be explained after the statement of the procedure.

1. Select all the nodes without a successor and arbitrarily label them from 1 to s, where s is the number of such nodes.* Each of W_{11}, W_{12}, and W_{13} is such a node, and we arbitrarily label them $W_{11}(1)$, $W_{12}(2)$, and $W_{13}(3)$ in the figure. All this says that in the schedule to be determined later, W_{11}, W_{12}, and W_{13} are assigned the ranks of 1, 2, and 3 that give them the three lowest scheduling priority among all the tasks.
2. Suppose that the first k tasks have already been labeled. For each of the remaining tasks all of whose successors have been labeled, make a decreasing list of the labels of their successors, and call it the

* There should be at least one such task since every task flow graph is a directed acyclic graph.

successor list of the task. Let $M = (m_1, m_2, \ldots, m_r)$ and $N = (n_1, n_2, \ldots, n_q)$ be any two such lists. Define an ordering "$\leq$" between N and M as follows: $N \leq M$ if (a) $m_1 < n_1$, or (b) for some i, $2 < i < q$, $m_j = n_j$ for all j, $1 < j < i-1$, and $m_i < n_i$ or (c) $q < r$ and $m_j = n_j$, $1 < j < q$. Among all the tasks, all of whose successor tasks have been labeled, determine the task whose successor list is the smallest among the successor list of all such tasks, and label it $k + 1$.

3. Repeat step 2 until all tasks are labeled.

Effectively the algorithm labels the tasks from the sink tasks (those without any successors) toward the source nodes (those without any predecessors). This way, tasks further away from the sink tasks are assigned higher labels.

4.2.5   Example

Let us apply these steps to the task flow graph in Figure 4.7a. We already labeled tasks $W_{11}(1)$, $W_{12}(2)$, and $W_{13}(3)$ in Step 1. During the next iteration of step 2, we see that W_9 and W_{10} are the only tasks all of whose successors have been labeled. Furthermore, their successor lists are* (1) and (3,2). By the ordering $\leq$ defined in step 2, (1) $\leq$ (3,2) so that W_9 is labeled $W_9(4)$, and since task W_{10} still remains as the only task all of whose successors are labeled, it is labeled $W_{10}(5)$. Repeating step 2 one more time, we find that W_7 and W_8 are the only tasks all of whose successors have been labeled and their successor lists are (5) and (5,4). Given that (5) $\leq$ (5,4), W_7 is labeled $W_7(6)$ and W_8 is labeled $W_8(7)$. During the next iteration of step 2, W_6 is labeled $W_6(8)$. Step 2 is iterated two more times to label $W_5(9)$ and $W_4(10)$. Finally, the tasks W_1, W_2, and W_3 are labeled $W_1(11)$, $W_2(12)$, and $W_3(13)$ in that order.

Once the labeling is completed, the next phase of the algorithm assigns the tasks to the two researchers using the following rule: Whenever a researcher becomes available, assign that researcher the task with the highest label and all of whose predecessors have been completed.[†] Following this rule results in the work flow and corresponding schedule in Figure 4.7b,c. Thus if we assume that each task takes 2 months to complete, the entire project takes 16 months to complete.

Coffman and Graham proved the optimality of this algorithm for any task flow graph and two processors. As noted earlier, in this example, the optimality of the algorithm is more directly seen by the fact that 13 tasks, each taking 2 months, cannot be completed in less than 14 months using

---

[*] Note that we use the new labels assigned to the tasks in the successor lists.

[†] If more than two or more researchers are available, arbitrarily pick one. In the example, we pick the research with the higher index.

Table 4.2 Work Flow Schedule Using Three Researchers

Time	2M	4M	6M	8M	10M	12M	14M
$R_1 \rightarrow$	W_1	W_4	W_5	W_6	W_7	W_{10}	W_{12}
R_2	W_2			W_8	W_9	W_{11}	W_{13}
R_3	W_3						

two processors. The dependence between tasks W_4 and W_5 adds an extra step (2 months) since these two tasks cannot be assigned to the two researchers at the same time.

The Coffman–Graham algorithm can be used with more researchers as well. However, its optimality cannot be guaranteed for work flow schedules with more than two researchers. In this example, using three researchers results in the work flow schedule given in Table 4.2, reducing the project completion time to 14 months. In this work flow schedule, the tasks $\{W_1, W_2, W_3\}$ are performed concurrently. This schedule is also optimal since 13 tasks, each taking 2 months, cannot be completed in less than 10 months using three researchers,* and the extra 4 months come from the dependences between tasks W_4 and W_5, and W_7 and W_{10}, as before. Furthermore, using more than three research assistants cannot reduce the project completion any further since the maximum number of tasks that can be carried out concurrently is at most three.

4.2.6   Optimal work flow schedules for task flow trees

The Coffman and Graham algorithm provides a provably optimal schedule for any task flow, but only if two processors are used. T. C. Hu discovered another optimal schedule.[†] His algorithm applies to any task flow with a directed acyclic tree topology and without any restriction on the number of processors as described next.

The main idea is to apply what is a called a topological sorting to the task flow graph to partition it into levels of tasks. The root of the tree forms level 1 and all predecessors of the root form level 2, their predecessors form level 3, and so on. An example of a rooted tree with six levels is shown in Figure 4.8. The levels are determined by starting out at the root and labeling the predecessors of the tasks at each level i by level $i + 1$, $i = 1$, 2, 3, 4, 5.

Once the levels are determined, the algorithm recursively selects and assigns the task at the highest level (farthest from the root) and all of

---

* In fact, the lower bound is $26/3 = 9$ months, but since each task is scheduled to last 2 months, we use 10 months.

[†] Hu, T. C. 1961. Parallel sequencing and assembly line problems. *Operations Research* 9:841–848.

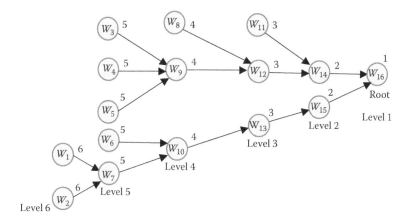

Figure 4.8 A directed tree, task flow graph and its levels.

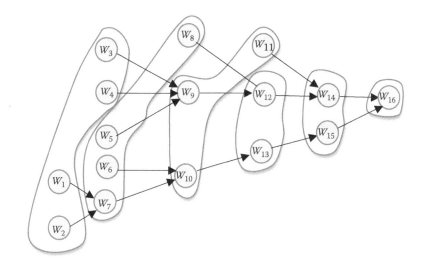

Figure 4.9 The work flow graph for the task flow tree using four processors.

whose predecessors have been assigned to the next available researcher. If several such tasks exist, then one of the tasks is arbitrarily selected and assigned, and the selection process is repeated until all tasks are assigned. If all researchers are synchronized, several tasks may be assigned to their researchers at once. The assignment given in Figure 4.9 demonstrates the process using four researchers. It is possible to combine tasks from different levels as long as there are no dependences between them as it happens between levels 6 and 5, 5 and 4, and 4 and 3. This work flow will take 12 months to complete if we assume that each task takes 2 months to perform. The fact that this matches the number of levels in the tree is related to the maximum number of tasks that can be performed concurrently and how the tasks are interconnected and leveled. In this case, using four researchers allows us to perform each level in one step.

In general, if the number of research assistants is greater than the maximum number of tasks in any given level of the task flow tree, then a work flow graph can be constructed to complete the project in mL months, where L is the number of levels in the task flow tree and each task is assumed to take m months. However, regardless of how many researchers are used, no work flow corresponding to a task flow with L levels can be completed in less than mL months with each task taking m months. This is because there must be at least one path of length L to have an L-level task flow tree, and all the tasks along such a path must be performed one after another, leading to mL months.

Hu and Coffman and Graham derived lower bounds on the minimum time and minimum number of processors needed to schedule an L-level tree of tasks. Here, we state the bounds given by Coffman and Graham for task flow trees in terms of number of months and number of researchers.

Remark 4.1 Let M denote the number of months needed to complete the work flow of an L-level task flow tree using r researchers with each task taking m months and let $q(j)$ denote the number of tasks located at levels greater than or equal to level j. Then

$$M \geq m \times \max_{1 \leq j \leq L} \left\{ j - 1 \left[\frac{q(j)}{r} \right] \right\}$$

Applying this lower bound (for $r = 4$, $m = 2$) to the task flow graph in Figure 4.8 yields the following minimum values for M:

j	1	2	3	4	5	6
$q(j)$	16	15	13	10	7	2
M	10	12	12	12	12	10

Thus, with four researchers, we need at least 12 months and this matches the workflow schedule given in Figure 4.8.

The second lower bound directly follows from the first and applies to the number of researchers needed to complete a task flow graph in M months.

Remark 4.2 Let M denote the number of months desired to complete the work flow of an L-level task flow tree using r researchers with each task taking m months and let $q(j)$ denote the number of tasks located at levels greater than or equal to level j. Then

$$r \geq \max_{1 \leq j \leq L} \left\{ \left\lceil \left\lceil \frac{q(j)}{\frac{M}{m} - j + 1} \right\rceil \right\rceil \right\}$$

Again, applying this lower bound (for $M = 12$, $m = 2$) to the task flow graph in Figure 4.8 yields the following minimum values for r as expected:

j	1	2	3	4	5	6
$q(j)$	16	15	13	10	7	2
r	3	4	4	4	4	2

Thus, if each task takes 2 months, 4 researchers are needed to complete the task flow in 12 months.

4.2.7 Work plan heuristics

The methods we have presented can be applied in certain circumstances to create work flows from task flows with optimal schedules. In general, developing an effective work plan to conduct a research project requires more than just converting task flows to work flows. Overall, the non-preemptive scheduling algorithms described here can be used for long-range schedules, and heuristics may have to be used to adjust such long-range plans as a research project progresses. However, the dynamics of a research project can easily foil a work plan sending the principal investigator back to the drawing board. To avoid costly setbacks and changes, it is important to have backup plans and use contingencies that may call for preemptive scheduling where resources may be preempted in real time. Preemptive scheduling algorithms are discussed in detail in the operating systems theory textbook by Coffman and Denning* and elsewhere.[†]

A good approach to increase the probability of reaching the goals of the project is to incorporate redundant steps into the task flows for alternative ideas, and derive optimal or near optimal work flows that include such redundancies. One practice that may be helpful in this regard is to assign research assistants a multitude of tasks without overloading them. This would not only generate the research capacity needed to incorporate the suggested redundancies into the project but also help research assistants have a broader research perspective and experience.

* Coffman, E. G, and Denning P. J. 1973. *Operating systems theory*. Prentice Hall.
† Blazewicz, J., Ecker, K. H., Pesch, E., Schmidt, G., and Weglarz, J. 2007. *Handbook on scheduling: From theory to applications*. Springer.

Making a work plan too rigid or structured as some of the previous examples suggest may be criticized on the basis that research projects should not be micromanaged and researchers would need some slack time to be creative. This is a valid criticism. However, as we stated at the beginning of the section, work plans serve to motivate researchers to work together and accomplish the goals of their projects. In the end, it is a delicate balance that principal investigators need to exercise to fine-tune the directions of their research projects to make sure that it remains on target while also taking some risks not to miss any golden opportunity that may come knocking.

4.2.8   Logging and tracking ideas

Working together in a research project with a research team requires planning and coordination of activities as in other projects, but with sufficient independence to ensure that the creativity of team members is not compromised. Synergy and independence must go hand in hand for a research project to succeed, and all researchers should be able to interact with one another and express themselves freely.

Research is a dynamic and liberal process, and ideas often get tossed around between the team members without getting recorded. It is valuable to maintain a secure electronic mail list where team members can compare notes and share their progress with one another while also depositing their new research ideas for further discussion. Sometimes creative ideas have to be revisited before they bear their fruits. Teams members may also keep their own diaries and record their findings and new ideas for future discussions.

In large projects, where it may be required to design and run experiments, build test beds and prototypes, and develop applications by teams of researchers, keeping things consistent and coherent can be a challenge. In particular, programming projects must have concurrent file processing and synchronization capabilities to allow several programmers to work together without breaking the integrity of the project.

4.3   Reaching research milestones and publicizing results

Establishing that a research project is progressing along and accomplishing its goals can be a challenge if it is not handled properly. Research is a competitive process, and every competitive researcher would like to make significant contributions and be the first one to discover the most significant results in his or her field. Often, the timing of the announcement of results is crucial to earning recognition.

History of science has a number of breathtaking examples of stiff competition among some of the greatest scientists. Evidently, one of these competitions took place between David Hilbert and Albert Einstein in 1915 on field equations describing general relativity*:

> Einstein presented his final equations on 26 November 1915, to the Prussian Academy. Hilbert's (1915) communication of a paper with the same equations to the Gottingen Academy is dated 20 November 1915. Hilbert was not quite as far ahead as a simple comparison of these two dates would suggest. This was shown by the recent discovery in the Gottingen archive of the proof pages of Hilbert's (1915) paper. These proof pages were revised substantially to yield the final published paper. In the unrevised version, the equations in a form comparable to those communicated by Einstein on 26 November are not in evidence. Nonetheless, Hilbert was extremely close.

Questions as to exactly what transpired in the written communications between Einstein and Hilbert and the content of Hilbert's original submission to the Gottingen academy in November 1915 remain unclear and are still disputed by historians of science. What is not disputable is that two of the greatest scientific minds competed and worked together[†] to build the most comprehensive theory to date in physics.

Similar disputes have been reported to happen between other great scientists like the discovery of differential calculus between Isaac Newton and Gottfried Leibniz, and the invention of the electronic computer between John Atanasoff, and John Eckert and John Mauchly. The latter ended in a court ruling in 1973 that decided that the ENIAC designed and built by Eckert and Mauchly in the Moore School of Engineering at University of Pennsylvania was derived from Atanasoff's ABC computer.[‡,§]

If these examples are to serve any purpose, several mechanisms are available to researchers to help document and publicize their results

---

* Norton, J. D. 2000. Nature is the realization of the simplest conceivable mathematical ideas: Einstein and the canon of mathematical simplicity. *Stud. Hist. Phil. Mod. Phys.* 31:135–170.

† Indeed, Hilbert invited Einstein to Gottingen for a talk in the summer of 1915 and Einstein stayed in Hilbert's house. See Rowe, D. E., 2001, Einstein meets Hilbert: At the crossroads of physics and mathematics, *Phys. Perspec.* 3:379–424.

‡ Burks, A. R. 2003. *Who invented the computer?:* The *legal battle that changed computing history.* Prometheus Books.

§ Mackintosh, A. R. 1987. The first electronic computer. *Phys. Today* 40:25.

while conducting a research project. With the extensive recording technologies of today, key findings during a research project can be documented and recorded electronically by taking pictures, and recording voice and video clips. In addition, universities and companies provide ample resources for researchers to disclose their intellectual contributions as technical reports and provisional patent applications. There are online databases such as arxiv.org that allow researchers to submit their work in progress in the form of preliminary manuscripts and even as fully developed papers.

There are also the usual platforms where the results from a research project can be publicized. These include conferences and scientific meetings, and archival journals. Conferences are ideal for rapid dissemination of ideas and results, but archival journal publications are a must in most fields of research for longevity of contributions. Of course, there are exceptions, and some researchers pride themselves on the success rates of getting their papers accepted by highly selective conferences. Publicizing research results through seminars and colloquium talks can also be an effective mechanism to popularize ideas and disseminate new results. It is important to be proactive and volunteer for talks when opportunities present themselves.

4.4   Reporting functions

One of the most critical functions of a funded research project is to report the findings of the project. Some funding agencies require periodic reporting to monitor and gauge the progress of research projects they fund. This may irritate some researchers. However, there is value in making researchers feel responsible as loosely monitored projects may cause investigators to drift away and lose touch with funding agencies. Worse yet, they may find themselves in a state with very few results to report at the end of the project if they do not set milestones to reach during the project.

One concern about periodic reporting that quite a few researchers would agree on is that presenting and publishing results takes a long time, and periodic reporting makes it difficult to accumulate enough presentations and publications in short periods. This argument may be more valid for some research projects than others, but periodic reporting templates typically include sections for reporting preliminary experimental findings and informal summaries of research results. Having a comprehensive work plan can also make it easier to write periodic reports.

Final reports are more exhaustive as they require funding agencies to meet their own goals and justify the use of funds. Such reporting requirements include describing contributions to the field and other fields, listing names of graduate students who have been funded by the project and the degrees they earned, listing journal publications and conference presentations, any outreach activities, and tools and computer applications and online Web sites developed during the project.

Different funding agencies use different templates but their expectations and requirements for a final report are similar. The following list of headings describes NSF's final report template:

- Project Participants
  - Senior personnel
  - Postdoctoral researchers
  - Graduate research assistants
  - Undergraduate students
  - Technicians, programmers
  - Other participants
  - Research Experience for Undergraduates (REU)
  - Organizational partners
  - Other collaborators and contacts
- Activities and Findings
  - Research and education activities
  - Training and development
  - Outreach activities
  - Journal publications
  - Books or other one-time publications
  - Web/Internet site
  - Other specific products
- Contributions
  - Contributions within discipline
  - Contributions to other disciplines
  - Contributions to human resource development
  - Contributions to resources for research and education
- Contributions beyond science and engineering

It is useful to know what funding agencies look for in final project reports as it can steer projects to match their outcomes with what is expected by funding agencies. Most of the information requested in the NSF final project report is straightforward, possibly with the exception of the contributions section. "Contributions within discipline" is asking for the new scientific and technical findings obtained in the project. "Contributions to other disciplines" include any cross disciplinary results that transcend the scientific discipline to which the project's subject belongs. This question typically applies to systems projects or those with an interdisciplinary focus. "Contributions to human resources" refers to an impact that has significantly affected the job preparation and skills of people involved in the project. "Contributions to resources for research and education" refers to results that impact the research infrastructure in the project's discipline and possible educational impact such as new curriculum materials, textbooks, laboratory facilities and tools that may enhance education at some level. "Contributions beyond science and engineering" refers to potential social implications of projects such as computer science research projects that focus on developing Web technologies that may impact large populations in social terms or research in other basic and applied sciences that may lead to products that affect lives and behavior of people.

4.5   *Working on several research projects at once*

Some investigators find themselves working on more than one research project at any point in time. If this happens out of the funding of sole investigator proposals, it represents a healthy state of affairs for investigators and indicates that their new ideas resonate with their peers in their field of research. Such simultaneous research projects may also have a synergetic effect and lead to results with broader implications within the investigator's field of research or even in adjoining fields.

However, working on a number of projects is often a consequence of submitting one-paragraph ideas to large-scale research proposals that involve a multitude of investigators. If investigators do not keep track of the number of one-paragraph ideas, they may find themselves working on half dozen or more projects at a time. Although this may feel good and look impressive on a résumé, getting involved in too many research projects all at once may be counterproductive to both the investigator and the projects in which he or she participates.

Unfortunately, universities have had very liberal policies in the past that led to a proliferation of research accounts supporting investigators without significant contributions or returns. Only recently, some restrictions are provisioned in submission requirements of funding agencies that aim to prevent researchers from signing up for research projects with one-paragraph ideas.

A good rule of thumb is that each new idea typically takes one research assistant or postdoctoral researcher to develop into a full-scale project working 20 hours a week. This means that an investigator needs to put in about 8 to 10 hours a week to actively pursue his new idea. A college professor teaching one course a semester typically spends 3 hours in class and with office hours and grading tests and assignments, and depending on the size of the class, may extend to 10 hours a week. With departmental and university service duties, it is reasonable to assume that he will easily be occupied 15 hours a week with teaching and service related work. This means that he or she will have time to supervise and manage no more than three research problems at a time. In the best of circumstances, this translates into three separate research projects. Having two sole investigator

research projects concurrently is more typical and should not be too difficult to handle. Of course, this scenario applies to a typical professor, and
those with administrative responsibilities will have less time for research
activities and should adjust their research load and time accordingly.

With these concerns notwithstanding, tackling certain research problems requires the research specialties of more than one investigator and is
best addressed by collaborative research proposals as long as each investigator adds a tangible and complementary expertise to the proposal. Such
proposals are written collectively by a number of investigators with one
or more of the investigators assuming an editorial role in composing the
ideas into a cohesive proposal. With the electronic sharing capabilities of
today, investigators need not be located in the same physical space and
can compose their ideas together using electronic mail and dozens of
online facilities that provide services ranging from video conferencing to
desktop sharing applications. Still, some collaborative research projects
with investigators working on campuses that may be located thousands
of miles away from one another can benefit from occasional face-to-face
meetings and interactions.

4.6   *Getting ready for next research proposal*

Every research project lasts only so long and investigators should plan
what comes next as their current projects begin to wind down. Typically,
a research project may end with one of three outcomes: None, some, or all
of the problems stated in the original research proposal have been satis-
factorily solved. In the rare likelihood that a research project fails to solve
any of the problems, it is not likely for its investigator to receive new funds
for another try. Likewise, if all the problems have been solved, there will
be very little incentive for any funding agency to fund an investigator to
solve the same problems one more time.

Thus, the only realistic possibility for a proposal to lead to another
proposal in the same field is if the investigator fails to solve some of the
problems he originally proposed or discover some new problems during
the current research project. Furthermore, there will be a higher prospect
of winning another award if the latter happened. For this reason, it is
important to use discretion when writing proposals. If too many problems are stated in the proposal and a large number of these problems
remain unsolved at the end of the project, the investigator will likely lose
credibility and reduce his or her chances of receiving additional funding.
On the other hand, proposing an easy set of problems may not impress
the reviewers or the funding agency and put a potential award at risk.

A reasonably safe approach for follow-on proposals is to propose (1) to
apply the new techniques developed during the current project to a different set of problems that are widely acknowledged by the key researchers

in the field and (2) to develop new techniques to solve the problems that surfaced during the current project. In each case, the amount of overlap between current project and new proposal is minimized as much as possible while some degree of continuity is established between them.

In general, funding decisions are made with the help of peer review panels, and peers are less likely to encourage an investigator to change or redefine their field of research. Peer review-based proposal evaluation systems are criticized exactly for this reason, and this will be discussed further in Chapter 6. To end the chapter on a positive note, it is safe to say that incremental research ideas are more likely to be funded by peer review-based systems, and to a large extent this is expected as there are very few fields where significant contributions are made on a daily, monthly, or annual basis. Yet, funding agencies across the globe receive tens of thousands of research proposals each year. This can only point in one direction, and that is, to put it mildly, most new ideas occur to more than one researcher and there are not many good ideas to go around.

4.7   Summary

This chapter reviewed the major activities that take place in funded research projects. It described how to assemble a research team, recruit research assistants, postdoctoral researchers, develop effective work plans, work flow schedules, and report and publicize research results.

4.8   Bibliographic notes

Project execution and reporting requirements of funding agencies are generally posted on their Web sites. In addition to annual and final reports, NSF also makes site visits to review project accomplishments. These and other project reporting requirements are described in an NSF document called the "Award and Administration Guide."*

4.9   Questions

  4.1 In the text, it was stated that competence, interest in the subject, and professional goals are three criteria that should determine if a graduate student qualifies for a research project. Are all of these three criteria equally important for you? Are there other criteria you would like to add to these or replace them with?
  4.2 Do you find it difficult to recruit qualified graduate students for your research projects? Would you rather work with postdoctoral researchers?

---

* National Science Foundation, Award and Administration Guide, http://www.nsf.gov/publications/pub_summ.jsp?ods_key=aag.

4.3 Would you have a need for undergraduate students in your research projects? What are the pros and cons of involving under-graduate students in your research?

4.4 How important is it for you to have a work plan to conduct a research project? Do you think that having a work plan will con-strain your research and limit your creativity?

4.5 For the following task flow graph obtain an optimal work flow using two researchers.

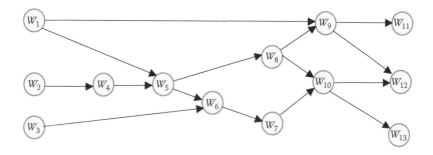

4.6 Determine the minimum number of months that would be needed to obtain a work flow with three researchers in the pre-ceding problem, assuming that each task takes two months to complete. Obtain an optimal work flow matching the number of months you have determined.

4.7 Convert the task flow graph in Question 4.5 into a task flow chart, assuming that each task takes two months to complete.

4.8 For the following task flow graph, use Hu's algorithm to obtain an optimal work flow schedule for three researchers. If each researcher works for two months, how many are needed to complete the project with four researchers? What is the short-est completion time for this task flow graph for any number of researchers?

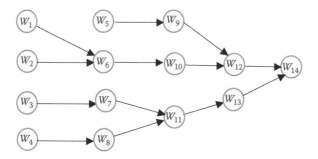

4.9 If your research project requires 15 tasks and each task takes 3 months how many research assistants would you need to complete your project in 30 months, assuming that all tasks can be assigned to research assistants?

4.10 Develop a work plan for a research proposal you plan to write, including a task flow and work flow. Use Remark 4.1 to determine the minimum number months you would need to complete the proposed research with your research assistants, postdoctoral scientists, and tools.

4.11 How often do you find others publish research results before you get to publish similar results of your own? Do you find presenting your ideas in conferences effective for publicizing your research results?

4.12 Do you find reporting requirements of funding agencies unreasonable? Do you find them helpful in conducting your research at all?

4.13 How would you characterize the contributions of your projects with respect to the classification of contributions given in Section 4.4? Would you change your research project to cover all classifications of contributions?

4.14 How often do you participate in large collaborative projects? What are the benefits and drawbacks of such projects for you?

4.15 How do you plan your next research proposal? How often do you switch to a new research subject after you complete your current research project?

chapter five

Promoting research in universities

Research and education are meshed together in the world's major research universities. Mission statements of such universities all emphasize the central role of scientific research in their undergraduate and graduate degree programs. Their campus construction and development projects, faculty hiring and firing policies, and student admission requirements all reflect this central role of research in their missions. In this chapter, the aim is to (a) reiterate the view that the very essence of research is all about learning while discovering and discovering while learning, (b) describe the responsibilities of administrative bodies and personnel for promoting research on their campuses, and (c) offer a set of quantifiable performance measures that can be applied uniformly to evaluate university researchers in a number of roles they assume in performing their duties. The chapter also describes some of the common mistakes universities can avoid to reduce the risk of inadvertently stifling the very research programs they set out to create and build.

5.1 Synergy between education and research

Not all higher education institutions share the same vision about incorporating research into their degree programs. However, they collectively serve the greater purpose of educating future generations of professionals for the betterment of the society at large, in all avenues of human endeavor, as shown in Figure 5.1. A two-year college strives to educate students for rudimentary postsecondary knowledge and skills, and prepare them for an undergraduate program at a more comprehensive college. Colleges that award predominantly bachelor's degrees measure their successes with the placement of their graduates in private sector and government jobs, and master's and doctoral programs in major research universities. Research universities measure their accomplishments by not only their graduates but also research grants and publications of their faculty members. They emphasize the impact of their research programs on the local and national economies through new scientific knowledge and technologies discovered in their research laboratories and licensed to companies. The new knowledge and technology generated at research universities

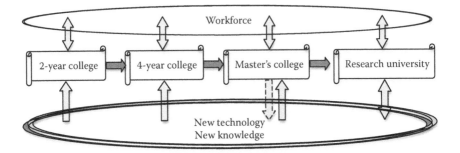

Figure 5.1 The cycle of research and education.

is returned to all educational institutions after being perfected into tools and products, including postsecondary colleges and universities. All colleges and universities contribute to the workforce in both directions, completing the cycle of research and education.

The Carnegie classifications* divide the U.S. postsecondary institutions into several categories based on a number of criteria such as undergraduate instruction, graduate program focus, program size and breadth, and enrollment profile. The categories of institutions in basic Carnegie classification, their numbers, and the degrees they awarded in 2009, are shown in Table 5.1. Research universities have been further classified into three categories based on the level of their research activities, in addition to the condition that each such university must award at least 20 doctorate degrees in the year they are evaluated. It is seen that 284 research universities of 3420 higher education institutions account nearly half of all bachelor's degrees, more than half of all master's degrees, and more than 85% of all doctoral degrees, while awarding less than 3.5% of associate degrees. Master's colleges and universities, which are classified as higher education institutions awarding fewer than 20 doctoral degrees and at least 50 master's degrees, provided more than one-third of all bachelor's degrees and more than 40% of all master's degrees. It should be noted that the 11,682 doctoral degrees awarded by the same category of institutions include 9010 doctor's degrees for professional practice, which are primarily graduate degrees in medicine, veterinary medicine, pharmacy, dentistry, and law. Doctoral degree figures for research universities also include sizeable numbers of professional doctoral degrees. Baccalaureate colleges, which are classified as higher education institutions awarding fewer than 50 master's degrees or 20 doctoral degrees, produced about 14% of all bachelor's degrees, less than 3.4% of all master's degrees, and

---

* Carnegie Foundation for the Advancement of Teaching, Carnegie Classifications, http://classifications.carnegiefoundation.org/descriptions/basic.php; WebCite, http://www.webcitation.org/5x20ub9Li.

Table 5.1 Carnegie Basic Classification Institutions and
Degrees Awarded in 2009

Institution Category	Number of Institutions	Associate's	Bachelor's	Master's	Doctoral
Research university– Very high activity	99	3010	383,194	164,300	51,163
Research activity–High activity	103	2834	255,501	98,609	15,460
Doctoral/ research universities	82	26,787	106,192	84,440	11,041
All doctoral universities	284	32,631	744887	347,349	77,664
Master's colleges and universities	654	29,694	582,426	255,001	11,682
Baccalaureate degree colleges	752	42,094	216,685	21,183	1,555
Associate degree colleges	1730	662,052	7331	1219	0
All institutions	3420	766,471	1,551,329	624,752	90,901

Note: The statistics were generated using IPEDS online tools from U.S. Department of Education, Institute of Education Sciences, National Center for Education Statistics, http://nces.ed.gov/ipeds/datacenter/Default.aspx.

a little over 1.5% of all doctoral degrees. Expectedly, associate's degree colleges awarded the lion's share of all associate's degrees with less than 0.5% of all bachelor's degrees, a practically insignificant number of master's degrees, and no doctoral degrees.

The basic Carnegie classification also includes some 757 special focus institutions that offer degrees in a single field such as medicine, law, music, and theology, and some 31 tribal colleges, which are part of the American Indian Higher Education Consortium.* The special focus institutions produced 17,258 doctoral, 36,242 master's, 58,578 bachelor's, and 24,333 associate degrees. More than 75% of the doctoral degrees were awarded in professional practice. Tribal colleges awarded 208 bachelor's degrees and 1409 associate degrees.

* The list does not include special focus institutions and tribal colleges.

The degree statistics in Table 5.1 demonstrate that the aspirants of undergraduate degrees place a higher premium at research universities. The mission statements of some of the best research universities reflect this affinity between research and education:

> "The synthesis of teaching and research is fundamental to Stanford University. Much of this faculty-driven research takes place in Stanford's schools, departments, and interdisciplinary programs."[*]
>
> "The mission of MIT is to advance knowledge and educate students in science, technology, and other areas of scholarship that will best serve the nation and the world in the 21st century."[†]
>
> "Cornell's mission is to discover, preserve, and disseminate knowledge; produce creative work; and promote a culture of broad inquiry throughout and beyond the Cornell community."[‡]
>
> "Harvard strives to create knowledge, to open the minds of students to that knowledge, and to enable students to take best advantage of their educational opportunities."[§]
>
> "Like all great research universities, Yale has a tripartite mission: to create, preserve, and disseminate knowledge."[¶]
>
> "The mission of the University of Cambridge is to contribute to society through the pursuit of education, learning, and research at the highest international levels of excellence."[**]

These ambitious missions are implemented by introducing imaginative learning and research activities into undergraduate programs. These include capstone design courses, honors programs, undergraduate thesis options, product-oriented research and development projects, and summer research internships. With the help of such activities, students develop skills to read and appreciate the contributions of established researchers, learn about recent advances in their fields of study, and actually carry out scientific research and disseminate their findings. They realize that research is the engine of scientific discovery and new knowledge, and

---

[*] Stanford University, Stanford research, http://www.stanford.edu/research/; WebCite, http://www.webcitation.org/5wzdBbMw5.

[†] Massachusetts Institute of Technology, Mission, http://web.mit.edu/mission.html; WebCite, http://www.webcitation.org/5wzdUmPdb.

[‡] Cornell University, The Cornell University mission, https://cornell.edu/about/mission/; WebCite, http://www.webcitation.org/5wzdYXlim.

[§] Harvard University, Mission statement, http://www.harvard.edu/siteguide/faqs/faq110.php; WebCite, http://www.webcitation.org/5wzdkn2EI.

[¶] Yale University, University mission statement, http://www.yale.edu/about/mission.html; WebCite, http://www.webcitation.org/5wzdpj84d.

[**] University of Cambridge, The university's mission and core values, http://www.admin.cam.ac.uk/univ/mission.html; WebCite, http://www.webcitation.org/5wzeJp32H.

without research, scientific progress will stall and there will be no new knowledge left to learn.

5.2 Graduate study and research programs

Compared to undergraduate programs, the synergy between teaching and research is built into graduate programs in postsecondary institutions much more naturally with comprehensive doctoral and master's degrees. Graduate study and research programs within basic instructional units or departments at a university constitute its research cores and perform its key research functions. Often, these programs are also embedded into or coupled with other university-wide centers and institutes designed to facilitate an administrative and logistical infrastructure for multidisciplinary research projects.

The key actors in a graduate research program are (a) regular faculty, (b) research faculty, (c) postdoctoral scientists and engineers, and (d) graduate (and possibly some advanced undergraduate) students. Collectively, they form the research engines within an instructional or research unit as shown in Figure 5.2. Regular faculty members submit research proposals, conduct research, serve as advisors to doctoral and masters students, publish and present scientific research results, and develop topics courses in their fields of research. Research faculty

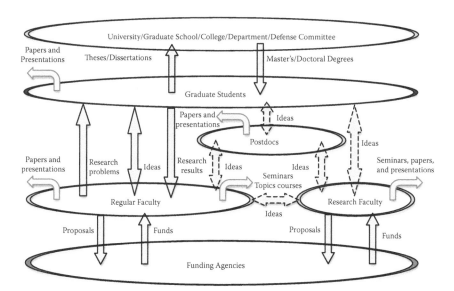

Figure 5.2 Graduate research activities in an instructional unit or department at a university.

members perform similar duties with the possible exception of serving as advisors in an official capacity to graduate students even though they interact with them. Postdoctoral scientists and engineers work with both regular and research faculty members and also interact with students. Graduate students serve as the buds of research ready to blossom with new ideas and produce new results. Besides these actors, what makes a graduate research program operate is the external funding that comes from funding agencies. Without external funds, universities will quickly find themselves turning into teaching institutions. This is what makes tenure and promotion decisions at research universities as contentious and difficult processes as they have come to be known. Faculty at research universities understand that they must raise their own funds to explore their research ideas, but to succeed, they need the support of their department chairs and college deans. Writing proposals takes a considerable amount of time and may not always lead to awards. Administrators often overlook this fact when they decide the teaching loads and other duties of their faculty.

To effectively run graduate programs, basic instructional units and departments are often organized into loosely partitioned groups of faculty by research areas. This partition makes it easier to assign faculty members to students as advisors and determine the courses they should teach. In general, teaching assignments for graduate courses should be made with two objectives in mind: (a) courses must be taught effectively and (b) faculty members involved in research must be able to remain in touch with graduate students. Satisfying both these constraints are complicated by the fact that effective teaching does not necessarily come with a high research portfolio or visibility. In fact, as discussed by Hattie and Marsh,[*] some would argue that there is an antagonistic relation between teaching and research as a faculty member has a fixed amount of energy and time to invest. Investing it in one more will inevitably deprive the other. Others think that there is a complementary and synergistic relation between the two activities, and this is the view that is promoted in the mission statements of research universities. It is difficult to find a scientific study that establishes a strong connection between teaching and research and some studies even suggest the absence of any such correlation.[†,‡] However, this does not mean that someone who is not

---

[*] Hattie, J., and Marsh, H. W. 2002. The relation between research productivity and teaching effectiveness: Complementary, antagonistic, or independent constructs. *J. Higher Educ.* 73:603–641.

[†] Shore, B. M., Pinker, S., and Bates, M. 1990. Research as a model for university teaching. *Higher Educ.* 19:21–35.

[‡] Hattie, J., and Marsh, H. W. 1996. The relationship between research and teaching: A meta-analysis. *Rev. Educ. Res.* 66:507–542.

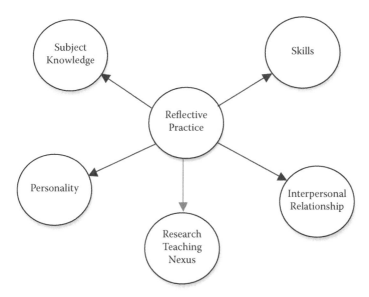

Figure 5.3 Kane et al.'s five-dimensional model of effective teaching. (From Kane, R., Sandretto, S., and Heath, C., 2004, An Investigation into Excellent Tertiary Teaching: Emphasizing Reflective Practice, *Higher Educ.* 47:283–310.)

familiar about a specialized research subject can deliver it as effectively as someone who is.

A more recent article also suggests a possible but a weak connection between research and teaching. It includes such a connection as one of five dimensions that influence effective postsecondary teaching.* These five attributes were derived from the academic profiles of a set of college science professors, recommended by their department heads as excellent teachers and their own descriptions of their teaching styles. The five attributes are (1) subject knowledge, (2) skills, (3) interpersonal relationship, (4) research/teaching nexus, and (5) personality. They are combined together through a hub that the authors call the reflective practice, as shown in Figure 5.3. The reflective practice explores how the five attributes of effective teaching interact with one another. The authors suggest that some of the professors polled in the study stated that they did make use of their research in their teaching. It also points out that other studies hint at a subtle and complex relation between teaching and research.

Whether there is a connection between teaching and research, Kane et al.'s five-dimensional reflective practice model suggests a possible solution

---

* Kane, R., Sandretto, S., and Heath, C. 2004. An investigation into excellent tertiary teaching: Emphasizing reflective practice. *Higher Educ.* 47:283–310.

to accomplishing both objectives of teaching assignments mentioned earlier. The model emphasizes personality and interpersonal relationships that could indirectly capture motivation for teaching. Thus, faculty members with a high level of research activity and who are not particularly keen on teaching may be assigned to teach more topics courses than their normal teaching loads would require. For example, if a faculty member is to teach one introductory graduate course, he or she may opt out of it with two topics, or one topics and one seminar course in return. Topics courses will likely motivate such a faculty member to be more focused on his or her teaching. Of course, making such preferential assignments may also induce some envy among faculty members that are assigned to introductory graduate courses with large enrollments. Nevertheless, doing so will likely limit the potential risk of ineffective delivery of core subjects to large numbers of students. Here, the right criterion is to ensure the most effective flow of knowledge between the various actors of a graduate research program, especially from faculty members to graduate students as they are on the receiving end of learning. It should also be pointed out that faculty members in an instructional unit or department serve as members of a team. It is not realistic to expect that all faculty members will perform equally well in all aspects of their professional responsibilities.

Besides judicious assignments of faculty to graduate courses, the vitality of a graduate research program depends on the enthusiastic participation of all of its key actors in the dissemination of research ideas and results within the program and university. Regular research presentations by faculty members and graduate students in the program may especially be valuable for synergizing research ideas and establishing new collaborative research projects. There is a general tendency among faculty researchers to keep their research to themselves within their home institutions. Although there are obvious reasons for focusing on the dissemination of results elsewhere, this should not mean a total isolation from colleagues with similar research interests. Sharing new research results would help graduate students and fellow faculty members to know that they are working alongside researchers who are renowned in their fields of research. Some universities attract more of such outstanding faculty than others, and this is often what makes their research programs so competitive and sets them apart from other universities.

5.3   Measuring faculty contributions and performance

Broadly speaking, universities use three metrics to assess and reward their faculty. They are research, teaching, and service. Research metrics are external research funding, journal publications, conference presentations, research monographs and other research-related publications, and citations to research publications. Teaching is measured mainly by student evaluations, and contributions and innovations that enhance the instructional experiences of students such as writing lecture notes and textbooks, setting up teaching laboratories, and designing laboratory projects and experiments. Service is measured in three categories: service to the (a) university, (b) public, and (c) profession.

Research, teaching, and service capture the key objectives of a research university, and faculty members are rewarded to the extent that they contribute to these three objectives. Administrators often use these metrics to push their faculty in directions that they think will accomplish the goals of the university. Typically, all faculty members are given a percentage increase for cost of living adjustments, and merit raises are left to the discretion of department heads. Merit raises cause much dispute among faculty as many of them think that their contributions are not adequately rewarded. There is good reason for this feeling as the criteria used by department heads are rarely revealed to the faculty and this makes the prediction of the impact of performance metrics on faculty salaries a research problem in and of itself. Indeed, several research studies have been reported on how various research and teaching metrics impact faculty salaries.[*,†,‡] Konrad and Pfeffer[§] state:

> We argue that the presence of uncertainty or ambiguity in the evaluation and salary-determination process is one factor that diminishes the strength of the relationship between productivity and pay. This uncertainty is generated when the evaluation of productive output is ambiguous and when the nature of the salary allocation process is such that it is hidden from examination.

They further state:

> We thus expect that the connection between salary and productivity will be stronger in the absence of uncertainty or ambiguity about the evaluation and

[*] Konrad, A. M., and Pfeffer, J. 1990. Do you get what you deserve? Factors affecting the relationship between productivity and pay. *Admin. Sci. Quart.* 35:258–285.

[†] Gomez-Mejia, L. R., and Balkin, D. B. 1992. Determinants of faculty pay: An agency theory perspective. *Acad. Manage. J.* 35:921–955.

[‡] Grofman, B. 2009. Determinants of political science faculty salaries at the University of California. *Polit. Sci. Poli.* 42:719–727.

[§] Konrad and Pfeffer, "Do you get what you deserve?"

> salary-determination process. Uncertainty in the
> criteria or dimensions for evaluating faculty pro-
> ductivity, uncertainty generated by an absence of
> social relations among faculty, and uncertainty pro-
> duced in relatively closed and secret administrative
> structures should diminish the connection between
> faculty salaries and research productivity.

To reduce the uncertainty of what counts in the minds of their faculty, universities would need to put in place concrete and transparent
policies for merit raises and use quantifiable metrics by which faculty
can estimate their salary raises based on their accomplishments. In this
regard, formulas work much better than long discussions in department heads' offices that typically lead to frustrations on all ends. As
observed in Arreola,[*] quantitative measures captured by formulas will
not necessarily make faculty evaluations objective but they will make
them uniformly measurable and allow some degree of control on their
subjectivity.

There have been a few studies that address the issue of faculty performance using quantitative measures. The approach proposed by Arreola[†]
provides a good starting point. It can be made more robust by further
emphasizing the quantitative aspect of the evaluation and making the
metrics of performance as objective as possible. For example, when measuring the teaching performance of faculty, using a student survey statistics or polling would be more reliable and objective than resorting to
isolated peer evaluations such as a class visit by a peer faculty or department chair's view of a faculty member's teaching skills. Such class visits
are often announced ahead of time and the faculty member under evaluation may be affected in one direction or another, leading to a less than
objective evaluation.

It can be stipulated that survey statistics gathered by student evaluations and other polling mechanisms may be subjective as well, but such
subjectivity can be controlled by a careful design of the survey questions and interpretation of the survey results. For example, students can
be asked if a professor has been very effective, effective, or not effective in teaching a class. This is not as polarizing as using a five-level
scale, where the teaching effectiveness is rated from excellent to poor
or from strongly agree to strongly disagree. Another reliable metric for
measuring the teaching effectiveness is the grade distribution of students
in certain service-oriented math, physics, and other courses. Teaching

---

[*] Arreola, R. A. 2000. *Developing a comprehensive faculty evaluation system*. Anker
Publishing.
[†] Arreola, *Developing a comprehensive faculty*.

effectiveness of faculty members can be compared in service courses where the same grading policies and curves are applied uniformly to all students in all sections. It is not unreasonable to assume that faculty members whose students perform better are more effective in delivering such service courses. If a peer assessment needs to be used in teaching evaluations, then it will be more objective to poll as many faculty members as possible.

Some assessments can be handled without using survey statistics. For example, the number of funded research projects or total funding provides a quantitative metric to judge a faculty member's funding performance without much dispute. Likewise, the number of articles published by faculty members and the numbers of citations they receive represent solid metrics of publication activity and impact in the field, and will be difficult to dispute.

With these views in mind, a multirole multifunction (MRMF) faculty evaluation model is proposed here as described in Arreola,* where faculty members assume multiple roles such as educators, researchers, administrators, committee members, thesis advisors, and mentors. In each of these roles, they can perform several different functions. For example, as educators they can teach courses, develop new courses, and write textbooks. As researchers, they can submit proposals and seek funding, write research articles, supervise graduate students, and attend and give talks at conferences and meetings. As part of their service role, they can serve as administrators, members of committees, editorial boards, conference program committees, and so on. These roles and functionalities are summarized in Table 5.2, and other roles and functions can be added as desired. For example, some universities encourage their faculty to be entrepreneurial and develop their research ideas into commercial technologies and products. This can be added as a new role with company founding, incubation efforts, fund raising, and other activities may be defined as expected functions. Of course, each of these roles must enhance the objectives of universities and it is only to that extent that universities will be eager to reward their faculties in each role.

Faculty members can be evaluated in each of these roles by developing appropriate metrics of performance for the corresponding functions. As we stated already, it is important to make these metrics as quantitative as possible to avoid criticism as much as possible. For research, all four functions listed in Table 5.2 can easily be quantified. For teaching, the total number of students taught, student evaluation scores, number of courses developed, number of books written, and number of undergraduate students advised can be used as quantitative measures. Similarly, for service, number of university committees served, number of profession-related

---

* Arreola, *Developing a comprehensive faculty.*

Table 5.2 Multirole Multifunction (MRMF) Faculty Evaluation Model

Role	Function			
Research	Seek funding	Publish papers	Present papers	Give talks
Teaching	Teach courses	Develop courses	Write books	Advise students
Service	Serve university	Serve profession	Serve community	Serve public

Source: Arreola, R. A., 2000, *Developing a Comprehensive Faculty Evaluation System,* Anker Publishing.

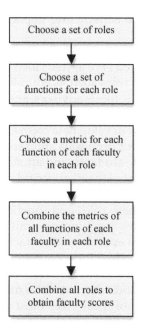

Figure 5.4 The multirole, multifunction faculty evaluation procedure.

positions held, and number of community and public events participated can be used as such measures.

The MRMF model can be used to develop a robust faculty evaluation procedure that focuses on a desired set of roles and functions. The steps involved in such a procedure are described in Figure 5.4. The procedure generates a set of scores for a pool of faculty for a given set of roles, each with an associated set of functions. The functions are combined together by a set of weights that specify the relative significance of each function. Similarly, the roles are combined by a set of weights that specify the relative significance of each role. Once faculty scores are computed, they can be converted to merit raises and other rewarding incentives.

In the remainder of this main section, we will focus on performance evaluation models to measure the research role of faculty and its associated functions. Each function in the research role evaluation can be measured quantitatively by its own metric, but it is also important to look at the bigger picture and assess the research roles of faculty members as a synergetic effort of interconnected functions. We will discuss both models as two alternatives for quantifying research performance of university faculty. Other faculty roles and functions can be assessed with the help of these models to obtain a broader faculty performance evaluation system as described in Figure 5.4.

5.3.1   The raw performance metrics model

In the first model, we use raw performance metrics such as the number of funded research projects, total funding amount received, number of graduate students, number of published research papers, and other similar factors to measure the relative research performance of faculty members. These raw performance metrics can be refined to differentiate between types of students (i.e., master's versus doctoral), impact factors of journals (e.g., class A versus class B journals), acceptance rates of conference papers (e.g., refereed or not refereed conferences), and so on.

To this end, consider a set of m faculty members, and let x_{ij} denote the raw performance of the ith faculty member for his or her jth function, $1 \leq j \leq n$. Let w_j; $1 \leq j \leq m$, be a set of values that add to 1. We define the joint performance, r_i, of the ith faculty by the following weighted arithmetic mean:

$$r_i = \sum_{j=1}^{n} x_{ij} w_j, i = 1, 2, \ldots, m,$$

(5.1)

where w_j; $1 \leq j \leq m$ are called the weights of the arithmetic mean.

Equivalently, in matrix form, we let

$$
\begin{bmatrix} r_1 \\ r_2 \\ \vdots \\ r_m \end{bmatrix} =
\begin{bmatrix} x_{11} & x_{12} & \cdots & x_{1n} \\ x_{21} & x_{22} & \cdots & x_{2n} \\ \vdots & \vdots & & \vdots \\ x_{m1} & x_{m2} & \cdots & x_{mn} \end{bmatrix}
\begin{bmatrix} w_1 \\ w_2 \\ \vdots \\ w_n \end{bmatrix}
$$

(5.2)

How the weights are chosen depends on the goals of the evaluation. Research universities place a considerable emphasis on the amount of research funding, number of research articles published, and number of citations to research articles, but other factors can be used for weighing in the raw performance metrics.

The joint performance metrics given in Equations 5.1 and 5.2 will work when raw performance values are close together. However, in most cases, such values are likely to span a wide range. For example, funding amounts are typically specified in hundreds of thousands of dollars, whereas the number of published papers can rarely exceed a dozen or two. This problem can be circumvented with the help of weights, but weights are designed to control the evaluation priorities between raw performance metrics. A better approach would be to normalize each raw performance value to the sum of raw performance values for all faculty members in the evaluated group. Accordingly, we define the normalized raw performance for the jth function of the ith faculty member by

$$\bar{x}_{ij} = \frac{x_{ij}}{\displaystyle\sum_{k=1}^{n} x_{ik}}, \tag{5.3}$$

where $0 \leq \bar{x}_{ij} \leq 1$, and $\sum_{i=1}^{m} \bar{x}_{ij} = 1; 1 \leq j \leq n$. The normalized raw performance metrics can be used to obtain the joint normalized performance, $\bar{r}_i$, of the ith faculty by replacing the x_{ij}'s by $\bar{x}_{ij}$'s in Equations 5.1 and 5.2:

$$\bar{r}_i = \sum_{j=1}^{n} \bar{x}_{ij} w_j; \ 1 \leq i \leq m \tag{5.4}$$

$$\begin{bmatrix} \bar{r}_1 \\ \bar{r}_2 \\ \vdots \\ \bar{r}_m \end{bmatrix} = \begin{bmatrix} \bar{x}_{11} & \bar{x}_{12} & \cdots & \bar{x}_{1n} \\ \bar{x}_{21} & \bar{x}_{22} & \cdots & \bar{x}_{2n} \\ \vdots & \vdots & & \vdots \\ \bar{x}_{m1} & \bar{x}_{m2} & \cdots & \bar{x}_{mn} \end{bmatrix} \begin{bmatrix} w_1 \\ w_2 \\ \vdots \\ w_n \end{bmatrix} \tag{5.5}$$

Given that $0 \leq \bar{x}_{ij} \leq 1; 1 \leq j \leq n$, and $0 \leq w_i \leq 1; 1 \leq i \leq n$, we have $0 \leq \bar{r}_i \leq 1; 1 \leq i \leq m$. Also, given that $\sum_{i=1}^{m} \bar{x}_{ij} = 1; 1 \leq j \leq n$, summing all $\bar{r}_i$'s shows that, if $\sum_{i=1}^{n} w_i = 1$, then $\sum_{i=1}^{n} \bar{r}_i = 1$. As the following example illustrates, this property is useful in converting joint normalized performances of a pool of faculty members to actual merit increases.

5.3.2 Example

Consider the group of eight faculty members whose raw and normalized raw performance values are shown in Table 5.3. The normalized performance values are computed from the raw performance values using Equation 5.3. Let $w_1 = \frac{1}{2}$, $w_2 = \frac{1}{4}$, and $w_3 = \frac{1}{4}$ denote the weights for funding, publications, and citations metrics, respectively. The joint normalized performances of the eight faculty, that is, $\bar{r}_i$; $1 \le i \le 8$, are computed using Equation 5.5 as shown in Equation 5.6.

$$
\begin{array}{c}
\text{Allen} \\
\text{Dung} \\
\text{Erkman} \\
\text{Gopalan} \\
\text{Hertz} \\
\text{Tales} \\
\text{Turtoise} \\
\text{Zellman}
\end{array}
\begin{bmatrix}
\bar{r}_1 \\
\bar{r}_2 \\
\bar{r}_3 \\
\bar{r}_4 \\
\bar{r}_5 \\
\bar{r}_6 \\
\bar{r}_7 \\
\bar{r}_8
\end{bmatrix}
=
\begin{bmatrix}
0.161 & 0.218 & 0.353 \\
0.080 & 0.127 & 0.142 \\
0.072 & 0.145 & 0.084 \\
0.082 & 0.218 & 0.026 \\
0.369 & 0.109 & 0.121 \\
0.051 & 0.055 & 0.089 \\
0.062 & 0.036 & 0.121 \\
0.123 & 0.091 & 0.063
\end{bmatrix}
\begin{bmatrix}
0.5 \\
0.25 \\
0.25
\end{bmatrix}
=
\begin{bmatrix}
0.223 \\
0.108 \\
0.093 \\
0.102 \\
0.242 \\
0.062 \\
0.070 \\
0.100
\end{bmatrix}
\qquad (5.6)
$$

Examining the table together with the joint normalized performance values of the faculty reveals that George Hertz and Elizabeth Allen, the faculty members with the highest most funding amounts, preserve their positions in the joint normalized performance ranking. However, Mark Zellman has been bumped from his third position in the raw performance ranking to the fifth position under the joint normalized performance evaluation. Both Weichen Dung and Kumar Gopalan improved their positions, whereas Alice Erkman, Andre Tales, and Jeanette Turtoise maintained their positions. This shows that even if the funding amount is given more weight in the joint normalized performance computation, it is possible to receive a lower ranking when one performs less than adequately in other functions.

It should be noted that the arithmetic mean of the joint normalized performances of any sample of m faculty members is $1/m$. This follows from the fact that $\sum_{i=1}^{n} \bar{r}_i = 1$. Thus, the mean of the joint normalized performances will tend to zero as m becomes large, effectively suggesting

Table 5.3 Research Funding Statistics for Eight Faculty Members

Name	Funding	Funding N.*	Publications	Publications N.*	Citations	Citations N.*
Elizabeth Allen	785000	0.161	12	0.218	67	0.353
Weichen Dung	390000	0.080	7	0.127	27	0.142
Alice Erkman	350000	0.072	8	0.145	16	0.084
Kumar Gopalan	400000	0.082	12	0.219	5	0.027
George Hertz	1800000	0.369	6	0.109	23	0.121
Andre Tales	250000	0.051	3	0.055	17	0.089
Jeanette Turtoise	300000	0.062	2	0.036	23	0.121
Mark Zellman	600000	0.123	5	0.091	12	0.063
Total	4875000	1.0	55	1.0	190	1.0

*N.: Normalized.

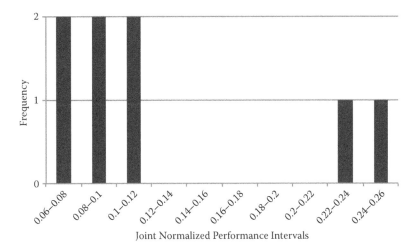

Figure 5.5 Histogram of joint normalized performances of eight faculty members.

that the frequency distribution of the joint normalized performances of any sample of *m* faculty members will have a positive skew for large *m*.* Consequently, nearly all faculty in the sample should have joint normalized performances near 0 as *m* gets large. The frequency distribution of the joint normalized performances computed in Equation 5.6 is shown in Figure 5.5 with an interval size of 0.02. The arithmetic mean of the joint normalized performances is $1/8 = 0.125$, and the standard deviation is 0.064. In this case, all joint normalized performances are less than 0.5, and they are split into two regions. In general, in any given distribution of joint normalized performances, only one may exceed 0.5, as they must all add up to 1.

The main challenge in making this metric work is how to choose the weight values in Equation 5.5. Like any other evaluation criterion, the joint normalized performance metric can be only as good as what it intends to measure. One way or another, university administrators should figure the values of weights, which are most acceptable to their faculty in judging their research performance.

---

* This can be established more formally using Mercer's inequality, $s^2 \leq 2(A - H)$, where s, A, and H denote the standard deviation, arithmetic mean, and harmonic mean of a set of numbers between 0 and 1 (see R. Sharma, 2008, Some more inequalities for arithmetic mean, harmonic mean and variance, *J. Math. Inequalities* 2:1009–114). Letting $A = 1/m$, and noting that $0 \leq H \leq A$ gives $s^2 \leq 2A = 2/m$. This shows that the variance of the joint normalized performances in Equation 5.5 is not more than $2/m$, which also tends to 0 as does the arithmetic mean, as *m* gets large.

5.3.3   Computing merit raises

Once the joint normalized performances are computed, they can be used directly as percentage increases in any merit pay since they add up to 100% as we mentioned earlier. For example, suppose that the department, where our eight faculty members work, has $50,000 for merit increases for research contributions. If we use their joint normalized performances as percentages of their merit increases, they would receive the merit increases shown in Table 5.4. It is seen that the lion share of the merit increases is almost evenly split between Elizabeth Allen and George Hertz even though Hertz's funding is more than double the amount that Allen brought to her institution. Others receive considerably less merit raises but given their funding levels and the compression that Hertz's funding has been subjected to, they will likely be content with the raises they receive.

5.3.4   Measuring effectiveness of faculty in conducting research

The raw research metric performance model that has just been described can be extended to include other research functions. However, it has one drawback in that it does not measure the effectiveness of faculty members in accomplishing their research functions. In the example given above, Hertz's $1,800,000 research funds resulted in only 6 publications, whereas Gopalan's $400,000 goes much further and produces 12 publications. Although a model focused on capturing the effectiveness of faculty members in making use of their research funds may not be a good approach to measure the faculty research performance, it can be valuable in assessing the effectiveness of using resources in research

Table 5.4 Merit Increases versus Average Research Metric for Eight Faculty Members

Name	Rank	Joint Normalized Performance	Merit Increase (in $)
Elizabeth Allen	2	0.223	11,150
Weichen Dung	3	0.108	5,400
Alice Erkman	6	0.093	4,650
Kumar Gopalan	4	0.102	5,100
George Hertz	1	0.242	12,100
Andre Tales	8	0.062	3,100
Jeanette Turtoise	7	0.070	3,500
Mark Zellman	5	0.100	5,000
		Total: 1.00	**Total: 50,000**

projects and guiding faculty to become more efficient in using their research funds.

The effectiveness of conducting research projects is a measure of the efficient utilization of various resources including research funds and faculty time. It can be quantified by associating with each research function of faculty members a pair of raw performance metrics that, in turn, quantifies their input and output performances for that function. For example, a faculty member may write a certain number of proposals, x, and receive a certain number of grants, y, in return. In this case, the hit rate or ratio, y/x can be used as a measure of the faculty member's effectiveness in using his or her time to convert proposals to actual grants. A faculty member who receives six grants by writing six proposals is more effective than the one who gets half as many grants for the same number of proposals. However, if we define the effectiveness as the ratio of the number of dollars received to the number of proposals submitted, the latter faculty may end up being more effective by receiving more dollars with three grants. Thus, effectiveness is not necessarily a better measure of performance than the joint normalized performance metric we described earlier. It is a measure that quantifies a ratio of two raw performance metrics of interest to determine how effectively one raw performance metric is converted into another one.

Some raw performance metrics may be indirectly related, and such indirect relations can be captured by a directed graph, called a research function tree, as shown in Figure 5.6. The x_i's represent the raw performance metrics that measure the activity levels of faculty members in their various research functions, whereas $e_{i,j}$'s capture their effectiveness

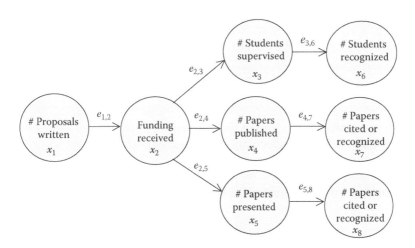

Figure 5.6 A research function tree of metrics and associated functions.

in performing these functions. For example, $e_{2,3}$ indicates how effectively research funds are converted to number of graduate students supervised. Similarly, $e_{4,7}$ quantifies the effectiveness of the number of published papers by the number of citations they receive during the evaluation period.

To compute the effectiveness between two indirectly related raw input and output performance metrics, we simply multiply the effectiveness values along the path from the input metric to the output metric. For example, in Equation 5.7, the effectiveness between the input metric x_1 (i.e., the number of proposals written by a faculty member) and the output metric x_8 (i.e., the number of citations made to his or her conference publications) can be determined as follows:

$$e_{1,8} = x_8/x_1 = (x_2/x_1) \times (x_5/x_2) \times (x_8/x_5) = e_{1,2} \times e_{2,5} \times e_{5,8} \qquad (5.7)$$

The effectiveness of a pool of faculty members for a given pair of input and output metrics can be determined from the effectiveness metric of the individual faculty members. Let $x_i; 1 \le i \le m$ and $y_i; 1 \le i \le m$ be the raw input and output performance values, respectively, for a pool of m faculty members. Let $Y = \sum_{i=1}^{m} y_i$. The effectiveness of the pool of m faculty, E, is given by the following weighted harmonic mean:

$$E = 1 \Bigg/ \sum_{i=1}^{m} \frac{\beta_i}{y_i / x_i}, \beta_i = y_i / Y \ ; \ 1 \le i \le m, \qquad (5.8)$$

where $\beta_i; 1 \le i \le m$ are called the weights of the harmonic mean. Simplifying the right-hand side of the formula after replacing β_i by y_i/Y shows that

$$E = Y \Big/ \sum_{i=1}^{m} x_i = Y/X, \qquad (5.9)$$

which is the ratio of the sum of the raw output performance values to the sum of the raw input performance values as expected.

Sometimes, it is more meaningful to work with a measure, which is the inverse of effectiveness. We call this measure the ineffectiveness metric, denote it by F, and define it as follows:

$$F = 1 \Bigg/ \sum_{i=1}^{m} \frac{\alpha_i}{x_i / y_i}, \alpha_i = x_i / X \ ; \ 1 \le i \le m, \qquad (5.10)$$

It is easy to establish the following relation between effectiveness and ineffectiveness metrics:

$$F = 1/E \qquad (5.11)$$

Table 5.5 lists the ineffectiveness values for the eight faculty members for funding amount, graduate student count, funding amount–publication count, and funding amount–citation count pairs of raw input and output metrics. For publication and citation counts, we have used the values in Table 5.3. For graduate student counts, we have used the list in Table 5.6. The average ineffectiveness values in the final column are computed by taking the weighted mean of the ineffectiveness values of the faculty members with weights $w_1 = \frac{1}{2}$, $w_2 = \frac{1}{4}$, and $w_3 = \frac{1}{4}$ in the order they appear in the table. Using Equation 5.10 or Equation 5.11, the harmonic mean of the three ineffectiveness metrics of the eight faculty members can be determined as 51, 861, 88, 636, and 25,657. The average of these ineffectiveness values with the same weights is 54,504. This is greater than the average ineffectiveness values of all faculty members except one. Higher values of the mean ineffectiveness imply worse performance. As the mean ineffectiveness values indicate, Hertz ends up being the most ineffective researcher since he spends $130,565.22 to fund each of his students, publish each of his papers, and a received a single citation with the given weights we have used. His raw performance values in student, paper, and citation counts make it impossible for him to rank higher in effectiveness. Even if the ineffectiveness values for the student and paper counts were completely ignored, he would still be ranked near the bottom of the list except for Gopalan. Despite his poor ineffectiveness values, it is difficult to think that Hertz will not be given a sizeable portion of the available merit dollars. Whether it would be as much as the amount, which was computed in Table 5.4 as the highest merit pay, will be determined by his university's priorities for research contributions.

5.3.5   Using citation indices to measure research impact

Citations to research publications are widely used in measuring the impact of research contributions of faculty researchers for promotion and tenure decisions as well as merit raises. For merit raises, department salary committees use the number of citations received by research publications of faculty members annually. For tenure and promotion decisions, a longer period of time, often the professional life span of faculty members under consideration, is used to collect a similar statistics.

A number of methods can be used to normalize the number of citations for a more objective evaluation of a pool of faculty members. One such method would be to use the effectiveness measure introduced in the previous section, and divide the number of citations by the total

Table 5.5 Ineffectiveness Indicators for Eight Faculty Members

Name	Funding	Funding/ Students	Funding/ Publications	Funding/ Citations	Mean Ineffectiveness
Elizabeth Allen	785000	49062.50	6416.67	11716.42	43814.52
Weichen Dung	390000	43333.33	55714.29	14444.44	39206.35
Alice Erkman	350000	43750.00	43750.00	21875.00	38281.25
Kumar Gopalan	400000	40000.00	33333.33	80000.00	48333.33
George Hertz	1800000	72000.00	300000.00	78260.87	130565.22
Andre Tales	250000	62500.00	83333.33	14705.88	55759.80
Jeanette Turtoise	300000	37500.00	150000.00	13043.48	59510.87
Mark Zellman	600000	42857.14	120000.00	50000.00	63928.57
Total	4875000	51861.70	88636.36	25657.89	54504.42

Table 5.6 Number of Graduate Student Advisees for Eight Faculty Members

E. Allen	W. Dung	A. Erkman	K. Gopalan	G. Hertz	A. Tales	J. Turtoise	M. Zellman
16	9	8	10	25	4	8	14

Table 5.7 Citation Impact Factors of a Group of Faculty

Name	Publications	Citations	Impact Factor
Elizabeth Allen	12	67	5.58
Weichen Dung	7	27	3.86
Alice Erkman	8	16	2.00
Kumar Gopalan	12	5	0.42
George Hertz	6	23	3.83
Andre Tales	3	17	5.67
Jeanette Turtoise	2	23	11.50
Mark Zellman	5	12	2.40

number of research publications of a faculty member. This ratio is often referred to as the impact factor. This metric may be used to measure the research impact of a pool of faculty researchers. In fact, the same metric is used to determine the impact factor of research journals as well.

As an example, the publication and citation counts of the group of eight faculty members described in Section 5.3.2 along with their impact factors are shown in Table 5.7. It is seen that Jeanette Turtoise has the highest impact factor even though she has only two publications and Andre Tales has the second highest impact factor with just three publications. The average number of citations may be partially misleading for measuring research impact as researchers with a small number of publications would likely have higher impact factors. This problem can be overcome by using what is called the Hirsch index or *h*-index, after its proposer Jorge Hirsch.[*] The idea of the *h*-index is to integrate the productivity and impact factor of a researcher into a single index. Thus, the *h*-index of a set of *n* research articles is defined as the integer *h* such that each of the *h* articles has at least *h* citations and each of the remaining $n - h$ articles have at most *h* citations.

It is relatively easy to compute the *h*-index of a set of research papers. All one needs to do is to list the papers by the number of citations they receive in descending order and scan the list down until the index of the paper from the top either matches or exceeds the number of citations it receives. If the index *i* of the paper matches the number of citations,

---

[*] Hirsch, J. E. 2005. An index to quantify an individual's scientific research output. *Physics,* ArXiv:0508025v2.

then i is the h-index of the paper. If the index exceeds the number of citations then $i - 1$ is the h-index of the paper. As an example, suppose that Elizabeth Allen's 67 citations are distributed among her 12 publications as follows:

Paper	1	2	3	4	5	6	7	8	9	10	11	12
# Citations	19	14	10	7	6	4	2	1	1	1	1	1

In this case, the sixth paper has the smallest index exceeding the number of citations it receives. Therefore, $i = 6$, and the h-index of Allen is $i - 1 = 5$. It is interesting to note that this almost matches the average number of citations of Allen. However, this is one of many distributions of 67 citations over 12 publications. It is possible to have some awkward distributions that may suggest misleading conclusions as in the following example:

Paper	1	2	3	4	5	6	7	8	9	10	11	12
# Citations	66	1	0	0	0	0	0	0	0	0	0	0

Here, the h-index drops to 1 as compared to the average number of citations of 5.58. In this case, the h-index practically ignores all but one of the publications, as they do not result in any tangible citations. This marginalizes the number of citations of Allen and makes h-index inadequate. Nonetheless, a well-known result, due to Erdos and Lehner,[*] can be used to establish that such cases rarely occur. This result states that the number of summands in almost all partitions of a positive integer n tends to $(3n/2)^{1/2} (\log n/\pi)$ for large n. For the example above, $n = 67$, and the Erdos–Lehner bound gives 13 to 14 summands. Thus, if we assume that all partitions of 67 citations are uniformly distributed, then there is an overwhelming probability that a randomly chosen partition will have 13 or 14 summands. Partitions such as $67 = 66 + 1$ or $67 = 66 + 2$ are not nearly as numerous as $67 = 7 + 6 + 5 + 5 + 5 + 5 + 5 + 5 + 5 + 5 + 5 + 5 + 4$. This makes distributions as in the second table example earlier highly unlikely, while making distributions like the one in the first table example much more likely.

Any set of research papers with an h-index of h must have at least h^2 citations between them since each of the h papers receives at least h citations. As pointed out by Hirsch, this lower bound is often not tight, as the citations to the remaining $n - h$ papers are not included in this count, and the number of citations to the papers with index values less than h is fixed to h. For the first example earlier, $h^2 = 25$ is much less than the actual

* Erdos, P., and Lehner, J. 1941. The distribution of the number of summands in the partitions of a positive integer. *Duke Math. J.* 8:335–345.

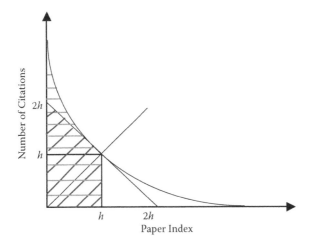

Figure 5.7 Estimating the *h*-index for a convex citation distribution. (The area marked by oblique lines forms a lower bound on the number of citaitons for a set of papers with an *h*-index of *h*.)

number of citations received by the papers published by Elizabeth Allen. For the second example, the lower bound estimate of h^2 papers becomes 1, making it even less useful. For convex citation distributions, a tighter lower bound can be derived by referring to the curve shown in Figure 5.7. The number of citations received by a set of papers is given by the area under the convex curve in the figure. The area marked with horizontal lines is the number of citations covered by the *h*-index. The area marked by oblique lines forms a part of this area, and thus provides a lower bound of $3h^2/2 =$ on the number of citations with *h* or more citations. For Allen, this gives about 38 citations as compared to her 67 citations. Hirsch suggests that the number of citations can be estimated as ah^2, where he found *a* to vary between 3 and 5, empirically.

The *h*-index has been very widely adopted by the bibliometric field and is likely to be around for a long time. Having said that, a large number of papers have been written on various shortcomings of the *h*-index. These include being unfair to authors with a small number of publications, not factoring the effect of multiauthor publications, not being uniform across different fields of research among others, and, most importantly, not taking into account any increases in number of citations to the highest most *h* papers.

A number of other bibliometric indicators have been suggested since the initial publication of *h*-index by Hirsch in 2005. Among these, the *g*-index introduced by Egghe* attempts to eliminate some of the

---

* Egghe, L. 2006. Theory and practice of the g-index. *Scientometrics* 69:131–152.

shortcomings of the h-index. The g-index of a set of research publications is a unique largest number g such that the g articles with the highest most citations collectively receive at least g^2 citations. Thus, the g-index averages the number of highest most g citations to g rather than lower bounding the number of citations among the highest h by h as in the case of the h-index. It is also true that $g > h$ for any citation distribution since h is the smallest index i that receives i^2 citations, and g is the highest such index.

Computing the g-index is not much more complicated than computing the h-index. We keep adding the number of citations starting with the highest number of citations until the running sum matches or becomes less than g^2, where g denotes the index of the last paper that is included in the sum of citations. For the two citation distributions given earlier in the section, the g-index values are computed as follows:

$$19 > 1^2, 19 + 14 = 33 > 2^2, 33 + 10 = 43 > 3^2, 43 + 7 = 50 > 4^2$$
$$50 + 6 = 56 > 5^2, 56 + 4 = 60 > 6^2, 6 + 2 = 62 > 7^2, 62 + 1 = 63 < 8^2$$

Paper	1	2	3	4	5	6	7	8	9	10	11	12
# Citations	19	14	10	7	6	4	2	1	1	1	1	1

Therefore, g-index = 7.

$$66 > 1^2, 66 + 1 = 67 > 2^2, 67 + 0 = 67 > 3^2, 67 > 4^2, ..., 67 > 7^2, 67 < 8^2$$

Paper	1	2	3	4	5	6	7	8	9	10	11	12
# Citations	66	1	0	0	0	0	0	0	0	0	0	0

Therefore, g-index = 7.

It is seen that, in both cases, the g-index is the same whereas the h-index values of the two distributions were computed as 5 and 1. In a way, the g-index is less discriminating against papers with fewer citations and in fact even with no citations. The other problem with the g-index is that it may tend to the number of publications for small sets of publications. For example, in the second table, if we truncate the number of papers to any number $x < 7$, the g-index will be equal to x. Effectively, this ignores the impact aspect of publications only giving credit to the number of publications.

The field of bibliometry is quite active and more research articles are being published to introduce more robust citation indicators (see, for example, the recent work by Hirsch and the citations given there*). With the emergence of such citation indicators, it is important to keep in mind that citation metrics are mere mathematical representations of the evaluation criteria we seek to apply. Thus, the *h*-index is useful if we wish to compare the citation records of a pool of faculty by the breadth of the impact of their papers. On the other hand, it is not at all useful if we are interested in the volume of their citations. Therefore, promotion and tenure, and other merit assessment bodies and committees at universities must match their priorities with such metrics before they use them in their assessments.

5.3.6   Using online bibliometric tools

Several online tools are available for computing the *h*-index of a set of publications.[†,‡,§] The Web of Science[SM] citation tool[¶] is easy to use as demonstrated next:

1. Using your university or company access privilege, connect to the Web of Knowledge[SM]. For universities, this access is generally provided through their main library Web sites. Check with your information technology office to learn how to connect to such a site at your university.
2. Once connected, type into the author field the name of the researcher for whom you wish to compute the *h*-index as shown in Figure 5.8a. At this point, you may also select or check off some options to limit your search to certain period of time or field of research such as science citation index, social sciences citation index, or arts and humanities citation index. In this example, we have entered "Allen ES" for Elizabeth Allen[**] and only included the science citation index with "all years" option. In most cases, specifying a last name together with two initials results in a fairly close match for the author being searched, but more work may be required as illustrated in the next step. Once the researcher's name and other options have been set, Web of Science returns the screen that is shown in Figure 5.8b, and where all the papers of Elizabeth Allen are listed.

---

* Hirsch, J. E. 2010. An index to quantify an individual's scientific research output that takes into account the effect of multiple coauthorship. ArXiv:0911.3144v2. F
† Citations-gadget, http://code.google.com/p/citations-gadget/.
‡ Web of Science, Citation report, http://images.isiknowledge.com/WOK45/help/WOS/h_citationrpt.html.
§ Scopus, Author evaluation tools, http://help.scopus.com/robo/projects/schelp/h_auteval.htm.
¶ Web of Science, Citation report, http://images.isiknowledge.com/WOK45/help/WOS/h_citationrpt.html.
** This is a fictitious name, and "Allen ES" does not correspond to a single person.

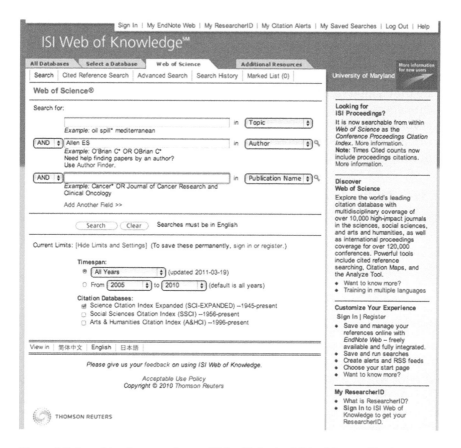

Figure 5.8 Searching for an author on Web of Science. (Web of Science^SM, Thompson Reuters, www.science.thompsonreuters.com.)

3. At this point, we can further restrict our options either using the "refine results" options on the left-hand side of the page or by clicking on the checkboxes next to the displayed articles. These are there to help ensure that we limit our search to the Elizabeth Allen we wish to search. Once this is done, we then click on the "Create Citation Report" link on the top right-hand corner as indicated by the transparent circle. This results in the display shown in Figure 5.9.

The plots on the left display the publications and citations of Elizabeth Allen between 1991 and 2010. The statistics on the right show her total number of citations, average citations per each of her publications, and her *h*-index. In this case, her *h*-index of 9 is considerably less than her average number of citations of 14.81. This would suggest that she has only a few highly cited papers as confirmed by the number of citations displayed.

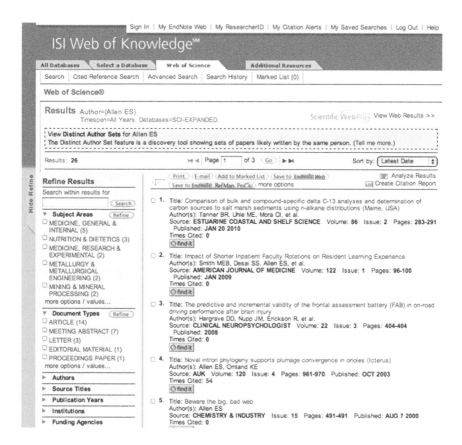

***Figure* 5.8** (Continued) Searching for an author on Web of Science. (Web of Science[SM], Thompson Reuters, www.science.thompsonreuters.com.)

The use of the *h*-index and other citation indicators is not limited to authors only. Any set of research articles that are generated by applying a desired set of search criteria, such as publications in a particular field, or in a particular journal, or those published by members of a faculty at a given department at a university can all be compared through such citation indicators. Table 5.8 lists the citation statistics for articles published by researchers affiliated with electrical engineering departments at some of the major research universities in the United States. The citation statistics were generated using the advanced search option in the Web of Science for years between 1981 and 2010. The departments have been ranked with respect to the three citation statistics that are provided by Web of Science as shown in the columns, which are labeled "Rank." The rightmost two columns (Pub C., Rank) list the numbers of papers published by the departments included in the statistics between

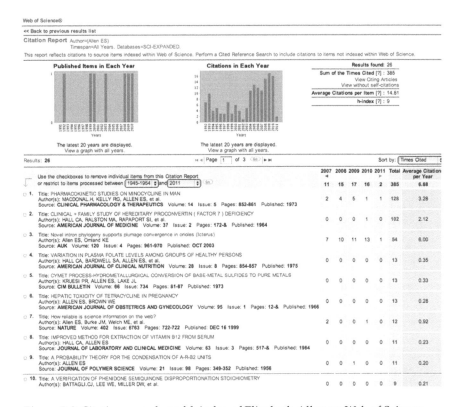

Figure 5.9 Citation records and *h*-index of Elizabeth Allen on Web of Science.

1981 and 2010 and the relative rankings of the departments. The entire table is ranked by the *h*-index values of the departments as well. All citations statistics include self-citations, but the Web of Science provides an option to exclude self-citations, if so desired. The column labeled "Citing R." lists the number of citations received by all the articles published by researchers affiliated with the corresponding universities. The entries in the impact factor column lists the average number of citations received by all the articles published by the people affiliated with the corresponding universities. The displayed citation statistics reveal some striking facts about their research impact. MIT unquestionably deserves its top ranking quantitatively in all three citation metrics. The citation statistics for other universities are also fairly consistent with one another for the most part with a few anomalies. For example, the California Institute of Technology has a very high impact factor, third after MIT and Princeton, but its *h*-index is surprisingly small as compared to other universities in the table. This can only be attributed to a more than usual number of very highly cited papers that would force the *h*-index to fall more rapidly. The

Table 5.8 Citation Statistics of Major Electrical Engineering Departments in the United States

University	h-Index	Rank	Impact Factor	Rank	Citations R.*	Rank	Pub. C.†	Rank
MIT	175	1	35.03	1	174,480	1	4981	3
UC-Berkeley	148	2	31.26	4	147,778	2	4727	4
Princeton U.	146	3	33.34	2	114,399	4	3431	8
U. Illinois-UC	130	5	21.19	7	133,304	3	6291	1
Stanford U.	125	4	24.46	5	93,549	5	3825	7
Purdue U.	119	6	21.17	8	90,948	6	4296	5
U. Michigan	112	7	18.13	15	90,389	7	4986	2
USC	110	8	21.27	6	70,952	8	3335	9
U. Maryland-CP	99	9	19.01	12	61,406	10	3230	10
Cornell University	97	10	21.17	9	60,788	11	2872	13
Georgia Inst. Tech.	96	11	15.45	20	64,494	9	4174	6
Penn State U.	94	12	19.33	11	52,199	14	2700	14
UC San Diego	89	13	17.85	16	55,828	12	3127	11
U. Texas-Austin	88	14	18.55	14	53,348	13	2876	12
U. Wisconsin	85	15	17.64	17	43,297	15	2455	15
California Inst. Tech.	83	16	32.99	3	34,479	19	1045	20
UCLA	82	17	18.71	13	40,027	16	2139	17
Columbia U.	81	18	20.64	10	32,382	20	1569	19
Carnegie-Mellon U.	75	19	16.14	19	34,835	18	2158	16
U. Washington, Seattle	73	20	17.37	18	35,344	17	2035	18

*R.: Received

†C.: Count

opposite inconsistencies arise between the *h*-indexes and impact factors of some other universities. For example, the University Michigan has a relatively low impact factor ranking as compared to its other rankings. The same is true for the Georgia Institute of Technology as well. These anomalies can be analyzed further by studying the distributions of citation counts in more detail.

## 5.4	Top five mistakes universities should avoid not to stifle research

Universities operate under a lot of pressure to balance their teaching and research functions. They serve clients with potentially opposing interests, and this may result in making poor judgment calls with respect to their research functions from time to time. The following list provides some of the most common mistakes that universities can avoid in order to not impair their research missions.

### 5.4.1	Mistake 1: Viewing research as an ordinary budget item

Administrators view the world through the prism of budgets they need to balance to manage their resources. Department heads often deal with veil threats from their deans and provosts that their budgets will be cut. They are told that they have to tap external sources if they want to maintain their support for conference expenses of their faculty and graduate students or upgrade their research laboratories. They are told that their faculty whose research funds dry up must be assigned more courses to teach. These are all the wrong signals and make faculty think that their institutions view research as a means for patching up budgets. With less resources allocated to covering the research activities of faculty and a growing perception that the priorities of a university are shifting toward other functions, departments will slowly drift toward mediocrity and lose their competitive researchers to other institutions. Here the right vision would be to create incentives for promoting research, for example, by returning a portion of the overhead charges to research projects, initiating university-wide faculty research funds, or using such overhead charges to provide emergency pools of funds for researchers who may experience temporary setbacks in their research funding.

### 5.4.2	Mistake 2: Failing to grasp the needs of a research university

Universities employ hundreds of faculty members who conduct research in their offices and laboratories on a daily basis. Managing the activities of such a large workforce of researchers is a formidable undertaking and requires a comprehensive infrastructure and administrative support. In most research universities, this role is assigned to an office of research administration and sponsored programs. As the size of such an office grows, the needs of faculty researchers can become more difficult to handle. This office must make sure that the administrative needs and priorities of all research projects, small and large, are met, and each individual researcher feels that his or her research program is as important and valuable as that of any other researcher. This is difficult to accomplish on a case-by-case basis, but policies that provide clear guidelines for supporting the infrastructural needs of researchers can go a long way to make them feel that they are being treated fairly and uniformly without regard to the content and size of their research projects. In addition, universities must make sure that their faculty and research scientists believe in their research philosophy and vision. Stating this is easier than doing so, as it requires the design and implementation of credible research policies that find broad approval and acceptance. Conducting regular surveys can help identify any issues that faculty experience in conducting their research projects, and reduce the distance between those who administer research funds and those who actually run the research projects at universities.

5.4.3   Mistake 3: Inadequate clerical support for processing proposals

Research funds provide universities with golden opportunities in more ways than one. To be sure, research projects carried out on a university campus bring credibility and prestige to that campus. But perhaps equally important for universities is that they receive large chunks of overhead dollars when their faculty win proposals and bring funds to their campuses. Many schools keep more than half of every research dollar brought to campus for what is called a "project overhead cost" or "facilities and administrative charge". It is true that the overhead moneys pay for electricity, water, heat, other utility expenses, and even the salaries of the key personnel that work in research administration offices of universities. However, after being charged that much overhead, nothing frustrates faculty members more than a terse announcement from a department head's office or university research administration office that states that they must complete their proposals a week before the submission deadline. It is obvious that administrative staff and managers feel the pressure of having to work on a large number of proposals just before a stream of deadlines, and paper trails take time to complete. Still, it helps if the paper trail process works as seamlessly as possible and does not pressure faculty to get things done way ahead of deadlines. This may seem unfair to research administration staff, but faculty members are constantly multitasking between a myriad of missions. Forcing them to shuffle their workloads will likely be counterproductive. University administrators from provosts down to department heads should realize that their administrative functions are designed to help faculty excel in every aspect of their work, including their research projects, and not to impose their will and ambitions on their busy schedules. Failing to do so will offend most faculty members and will give them an excuse not to carry out their research functions adequately. Of course, the best approach to streamlining the proposal submission process would be to use an all-electronic system with minimum human intervention. Both universities and funding agencies are moving their proposal and grant processing systems in this direction to minimize the potential congestions between research administration staff and faculty.

5.4.4 Mistake 4: Missing the cue on how to reward research

A common concern for most faculty members working at a research university is that they are not given a precise account of how their research efforts contribute to their professional standing and salaries within their departments and colleges. Most universities publish merit pay policies for their employees, but these policies provide general guidelines and fall short on specifics. Such policies should promote merit pay increases as a measure of rewarding faculty for serving the mission of the university and facilitating a dynamic interaction between administrators and faculty to help the university reach its educational and research goals. In particular, the determination of merit raises should be decoupled as much as possible from peer evaluations within the same unit as such evaluations tend to contain subjective views of faculty about the value of contributions of their colleagues. At the same time, rigid merit raise procedures tend to discourage faculty for putting themselves up through a long and drawn-out process. An ideal system should possibly involve an electronic system with a clearly defined merit point scheme where faculty can enter their contributions and be notified electronically of their total merit points, which can then be converted into actual merit raises through the payroll systems of universities.

Faculty may meet with their department heads and college deans to help resolve their issues and look for answers. However, nothing can substitute for a precise policy that declares—in as much quantitative terms as possible—what faculty members should expect to receive for the research dollars that they bring in, articles they publish, conference papers they present, and graduate students they supervise. Not having such a policy often demoralizes faculty and can make them lose interest in their research programs. The problem becomes more acute when faculty salaries begin to spread out. Any competitive faculty member who feels that his or her research efforts are not appreciated or less appreciated than those of others will have enough excuse to look elsewhere and leave as soon as an opportunity arises.

5.4.5   Mistake 5: Favoritism for research resources

It is true that administrators receive and return favors, and some would openly engage with their faculty members on a *quid pro quo* or favor-for-favor basis. Under normal circumstances and when applied uniformly, this approach may get things done and all will be good and groovy. However, a more selective application of this approach may generate a rift among the faculty and create an inhospitable environment where some members may feel that there is favoritism at work and their research projects are not given as high a priority as some other projects. Consider a situation where a group of faculty work collectively on research projects and have acquired some laboratory space in a department for conducting their experiments. After a while, another faculty member needs a similar laboratory facility but is not permitted to use that laboratory space by the group of faculty that controls it. At first, it may seem that the problem is trivial and a department head can simply provide the other faculty member with another space. What if there is no such space or the laboratory facilities needed by the lone faculty member are very expensive to acquire. Furthermore, what if the laboratory facilities used by the group of faculty members belong to the department? Ordinarily, the department head should have sufficient administrative privileges to intervene and make sure that the lone faculty member is able to conduct experiments using those facilities. However, sometimes this leads to a power struggle, especially if the lone faculty member is not tenured. This is one of many examples of turf battles that occur on university campuses all the time. It not only happens in the context of laboratory facilities but also other resources, including the allocations of teaching assistants as thesis advisees to faculty, impartial allocations of travel funds to attend meeting conferences, and so on. In resolving such conflicts, the best position that a university should take is that university-owned resources and facilities should be available for any researcher and shared on an equitable and fair basis. Other resources acquired by external research funding by individual faculty members can be shared on a courtesy and availability basis. Any other method that applies undue pressure without a rational basis argument will frustrate one or another group of faculty and further inflame the conflict.

5.5   Summary

This chapter described the key factors that underpin the success of a university in performing its research mission along with its education mission. Specific methods were presented with which this success can be quantified in terms of original scientific contributions of university researchers as well as how to stimulate research using equitable merit incentives and raises.

5.6   Bibliographic notes

The statistics in Table 5.1 were compiled using the Integrated Postsecondary Education Data System (IPEDS) online tools of the U.S. Department of Education, Institute of Education Sciences, National Center for Education Statistics. The statistics in Table 5.8 were compiled using the Web of Science online tools and database.

5.7   Questions

5.1 How do you view your role as a researcher in a university setting? Does this role enhance your teaching portfolio and make you a more effective teacher?

5.2 Do you view yourself more as a teacher or researcher? Which one would you prefer if you had to choose between them?

5.3 What is a research university in your opinion? Is classifying universities as in the Carnegie Classification appropriate? Would such a classification have a negative impact on researchers who work in undergraduate colleges?

5.4 How applicable is the five-dimensional model of teaching effectiveness described in the chapter to you? Are there other factors that determine your teaching effectiveness?

5.5 How applicable is the multirole multifunction model to measuring your job performance as a university professor? Should other roles and functions be added?

5.6 Do you feel that your research contributions are sufficiently recognized by your university? Would you prefer a separate evaluation of your research contributions in determining your merit pay?

5.7 How would you do things differently to promote research in your department?

5.8 Consider a pool of 10 faculty members, f_i and $1 \leq i \leq 10$ with their annual raw performance figures for the three research functions they serve as shown next. The three functions $g, p,$ and d represent the total grants received, journal articles published, and doctoral dissertations supervised.

Faculty	f_1	f_2	f_3	f_4	f_5	f_6	f_7	f_8	f_9	f_{10}
g	5,000	5,000	15,000	25,000	150,000	25,000	25,000	100,000	50,000	600,000
p	3	1	3	2	4	2	1	8	2	6
d	1	1	1	2	2	1	1	2	2	4

a. Compute the joint performance of each faculty for the three research functions they perform, using the weights $w_g = 1/50$, $w_j = 12/25$, $w_p = 1/2$, and Equation 5.1. Could you guess which faculty member would perform best without first computing their joint performances for the three functions?

b. How would you use the joint normalized performance values you have determined in (a) to allocate merit payments for this pool of faculty using a $100,000 merit allocation?

5.9 (a) Use the raw performance values given in the previous question together with Equations 5.3 and 5.4 to compute the joint normalized performances of the pool of 10 faculty members, and show that they sum to 1. (b) Use the joint performance values you have computed to determine an appropriate merit raise for each faculty member with a merit budget of $100,000.

5.10 What do the mean ineffectiveness values in Table 5.6 represent? How can effectiveness or ineffectiveness be used in motivating faculty for research?

5.11 Use Web of Science to plot the citations to your work and compute your *h*-index and *g*-index. Which of these indices do you favor, and why? Would you rather use the impact factor instead?

5.12 Compare your citation statistics with two fellow researchers and comment on any similarities and/or differences.

5.13 Compute the citation statistics of your department and compare it to those given in Table 5.8.

5.14 Many research universities are considerably invested in research and technology commercialization projects. Do you think that such activities are productive and help the research mission of universities?

5.15 We listed a number of mistakes universities make to adversely affect their research programs. What other mistakes do universities make, in your opinion? How do they affect your performance?

chapter six

Scientific research
A funding agency perspective

Scientific research is the bedrock in which new technologies incubate, and government and private funding is what fuels scientific research around the globe. The United States, Japan, and a few other countries that understand this key relationship have been investing tens of billions of dollars in scientific research annually to maintain their leadership in science and technology. The research and development expenditures of all countries combined exceeded $1 trillion dollars in 2007 with an upward trend.[*] With about 5.7 million researchers around the globe,[†] this corresponds to about $175,000 of funding per researcher per year. In 2008, U.S. universities spent $51 billion in research expenditures,[‡] and with nearly 275,000 researchers in U.S. universities,[§] this amounts to about $185,000 of funding per researcher per year. These numbers look impressive, but only a small fraction of all proposals submitted to funding agencies is supported. Although it can be said that competition serves as a filter for quality, it is also conceivable that, with the declination of overwhelming numbers of proposals, some potentially very creative ideas are lost forever. This brings up the question if funding agencies can be more resourceful and clever to increase their chances of investing on the right set of proposals. To this end, the aim in this chapter is to (a) describe how a model funding agency should be organized and function, (b) discuss how researchers can build a dynamic interaction with funding agencies to get their ideas across better, and (c) suggest proposal assessment and selection ideas that can be used to make the proposal funding process more effective with limited resources.

6.1 A funding agency model for scientific research

Like any organization, an agency that funds scientific research can be characterized by a mission and a body of functions that are designed to realize its mission. The mission of such an agency is typically specified

---

[*] NSF 2010 Science Indicators, p. O-4.
[†] NSF 2010 Science Indicators, p. O-8.
[‡] NSF 2010 Science Indicators, p. 4-4.
[§] NSF 2010 Science Indicators, p. 5-19.

by an act or charter that is enacted or mandated by the authority and endorsement of an elected or governing body such as a national assembly of citizens, state or federal government, or a private enterprise under which it is constituted. For example, the National Science Foundation (NSF) was established in 1950 under U.S. Public Law 81-507 as a federal government agency to fund basic scientific research on behalf of the U.S. government.* The NSF's mission was declared in the preamble of its act: "To promote the progress of science; to advance the national health, prosperity, and welfare; to secure the national defense; and for other purposes."

The NSF's mission statement is fairly broad in its scope: It ties scientific progress to national health, prosperity, and welfare and even includes a phrase at the end that provides for possible shifts of research focus over time. Another funding agency for scientific research, the Scientific Foundation of Ireland (SFI), which was constituted by the 2003 Industrial Development Act of Irish Parliament, posts the following mission statement on its Web site:† "SFI will build and strengthen scientific and engineering research and its infrastructure in the areas of greatest strategic value to Ireland's long-term competitiveness and development."

Still another funding agency, the National Research Council of Italy states its mission with a similar focus:‡ "To carry out, promote, spread, transfer and improve research activities in the main sectors of knowledge growth and of its applications for the scientific, technological, economic and social development of the Country."

All these mission statements can be reduced to a succinct representation: There is a positive correlation between human wealth and welfare and scientific progress. This empirically tested and validated proposition is what pushes countries to invest in scientific research around the globe. Beyond that, the specific functions to be performed to accomplish a funding mission concentrated in certain fields of scientific research or certain forms of funding are determined by the resources and priorities of a funding agency. For example, the central functions of NSF are stated in Section 1862 in US Code 42§:

(1) To initiate and support basic scientific research and programs to strengthen scientific research potential and science education programs

---

* Public law 507-81st Congress. Ch. 171-2D session.
† Science Foundation Ireland, Vision and mission statement, http://www.sfi.ie/about/vision-mission-statement/; WebCite, http://www.webcitation.org/5xK8HCdYO.
‡ CNR, About CNR, http://www.cnr.it/sitocnr/Englishversion/CNR/AboutCNR/About CNR.html; WebCite, http://www.webcitation.org/5xKJdEw7r.
§ Cornell University Law School, U.S. Code § 1862 Functions, http://www.law.cornell.edu/uscode/42/usc_sec_42_00001862----000-.html; WebCite, http://www.webcitation.org/5xKoKUMOx.

at all levels in the mathematical, physical, medical, biological, social, and other sciences, and to initiate and support research fundamental to the engineering process and programs to strengthen engineering research potential and engineering education programs at all levels in the various fields of engineering, by making contracts or other arrangements (including grants, loans, and other forms of assistance) to support such scientific, engineering, and educational activities and to appraise the impact of research upon industrial development and upon the general welfare;

(2) To award scholarships and graduate fellowships for study and research in the sciences or in engineering;

(3) To foster the interchange of scientific and engineering information among scientists and engineers in the United States and foreign countries;

(4) To foster and support the development and use of computer and other scientific and engineering methods and technologies, primarily for research and education in the sciences and engineering;

(5) To evaluate the status and needs of the various sciences and fields of engineering as evidenced by programs, projects, and studies undertaken by agencies of the Federal Government, by individuals, and by public and private research groups, employing by grant or contract such consulting services as it may deem necessary for the purpose of such evaluations; and to take into consideration the results of such evaluations in correlating the research and educational programs undertaken or supported by the Foundation with programs, projects, and studies undertaken by agencies of the Federal Government, by individuals, and by public and private research groups;

(6) To provide a central clearinghouse for the collection, interpretation, and analysis of data on scientific and engineering resources and to provide a source of information for policy formulation by other agencies of the Federal Government;

(7) To initiate and maintain a program for the determination of the total amount of money for scientific and engineering research, including money allocated for the construction of the facilities wherein such research is conducted, received by each educational institution and appropriate nonprofit organization in the United States, by grant, contract, or other arrangement from agencies of the Federal Government, and to report annually thereon to the President and the Congress;

(8) To take a leading role in fostering and supporting research and education activities to improve the security of networked information systems.

These codified functions of NSF underline its central role in designing and implementing the research and development mission and

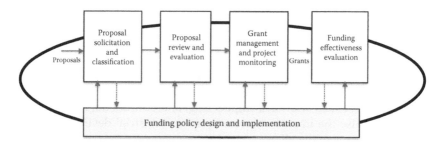

Figure 6.1 Basic functions of a funding agency for scientific research.

policies of the U.S. government and go much further than a funding operation alone. The skeleton of a scientific research agency that is focused primarily on funding, and its most basic functions are shown in Figure 6.1. The funding policies of a scientific research agency define its funding principles and serve as its operational guide. The other four functions deal with proposal processing and can be viewed as a four-stage pipeline. Proposals enter this pipeline through the proposal solicitation and classification stage, and exit it through one of the remaining three stages. All proposals submitted to the system go through the first two stages. Some reach the third stage for an award, and some or all of those may be subject to the funding effectiveness evaluation stage. Funding policies affect all four functions, and based on how these functions perform over time, they may be revised and result in revisions in any of the functions in return. The funding policies may also be modified by changes in an agency's mission due to government decisions and priorities.

Funding policies spell out the general funding requirements for proposals. In particular, the following questions are some of the key determinants of the funding policies of a scientific research funding agency:

- In what fields of research will the agency solicit proposals?
- What conditions, if any, will be imposed on proposal submissions and their investigators?
- What types of funding, investigator's time, equipment, travel, and so forth will be supported by the agency?
- How long will projects be allowed to run?
- Will there be any provisions for the distribution of funds by any demographic markers, for example, region, university affiliation, gender, and minority status?
- Will proposals with international investigators be funded? If so, will there be any restrictions on funding levels or types of funding?
- Will proposals be evaluated by peer review? What metrics will be used to evaluate them?

Each of these and possibly other similar questions need to be answered for a funding agency to operate and perform its functions within the bounds of its mission. For example, NSF supports research in the following broad fields of science and engineering:

- Biological sciences
- Computer and information science and engineering
- Engineering
- Geosciences
- Mathematics and physical sciences
- Social, behavioral, and economic sciences
- Education and human resources

These areas are organized as directorates, each headed by an assistant director. In addition, NSF has a number of other offices that operate directly under its director and handle international research programs, cyber infrastructure projects, polar research programs, integrative activities, and administrative and legal aspects of the foundation.*

The funding decisions of NSF are mainly made at the division level within the seven directorates and in terms of the various research programs established within each division using a peer review system. For example, the Mathematical and Physical Sciences Directorate consists of the following divisions:

- Astronomical sciences
- Chemistry
- Material research
- Mathematical sciences
- Physics
- Office of multidisciplinary activities

Each division is further divided into programs or program clusters. For example, the Division of Mathematical Sciences consists of the following research programs†:

- Algebra, number theory, and combinatorics
- Analysis
- Applied mathematics
- Computational mathematics

---

* Other aspects of funding that address the questions we raised are described in detail in NSF's *Grant Proposal Guide*.
† NSF, Mathematical sciences (DMS), http://nsf.gov/div/index.jsp?div=DMS; WebCite, http://www.webcitation.org/5xLUcnVHd.

- Foundations
- Geometric analysis
- Mathematical biology
- Probability
- Statistics
- Topology

In addition to these standard research programs, each division has a number of other funding capabilities that support conference, workshop, and training activities within its scope as well as cross-cutting programs to fund interdisciplinary research projects.

The examples illustrate that NSF uses a hierarchical (three-level tree) structure to classify research fields into focus programs. If a funding agency has a more limited set of funding areas, a two-level tree classification may be used.

6.2   *Proposal solicitation and submission process*

Funding agencies solicit proposals with established annual or biannual deadlines. Typically, each program solicitation has its own subject focus and submission requirements even though clusters of programs may also be involved in joint solicitations. It is important to read the requirements of a research program before submitting a proposal to that program. For example, the Geometric Analysis program in the Division of Mathematical Sciences at NSF has the following solicitation information:*

> The program in "Geometric Analysis" supports research on differential geometry and its relation to partial differential equations and variational principles; aspects of global analysis, including the differential geometry of complex manifolds and geometric Lie group theory; geometric methods in modern mathematical physics; and geometry of convex sets, integral geometry, and related geometric topics.

To be competitive, proposals submitted to this program would expectedly have to deal with research problems in one of the highlighted fields of research or a closely related field. If the synopsis of research fields in a program does not provide a clear direction for suitability of a submission to that program, checking out recently funded proposals in that

---

* NSF, Geometric analysis, http://nsf.gov/funding/pgm_summ.jsp?pims_id=5549&org=
  DMS&from=home; WebCite, http://www.webcitation.org/5xLUnl943.

program may provide additional insight. Most NSF programs provide a link to such proposals on their Web pages. The same information can be obtained using the "Awards" link at NSF's main Web site.

A number of proposals are declined as a result of being sent to the wrong programs or divisions as focus areas shift within a funding agency over time. It is crucial for principal investigators to make sure that their proposals are submitted to the right division and program to avoid such undesirable consequences. Studying the synopsis of programs and tracking the proposals that are funded are a good first step to determine where to submit a proposal, but this may not be sufficient in all situations. As an investigator, you may need to probe the priorities and inclinations of a program director in a face-to-face meeting, since program directors are the ones who pick the reviewers and make the crucial recommendations for funding. In an ideal world, program directors are expected to be subject-neutral, but the real world is hardly ideal, and program directors have their own favorite subjects. Therefore, if you have any hesitation for submitting your proposal to a particular program, it would help to meet with the director of the program and describe your research ideas. This is especially important if you are switching to another field of research and your work has not been fully recognized in the field. "I am not familiar with his work, therefore it cannot be that good" syndrome is widespread among program directors and reviewers, so beware.

A good preparation is essential before meeting with a program director. A one-page summary of your ideas and their importance can go a long way to familiarize a program director about your research. Depending on the format of the meeting, you should describe your ideas at a level of technical depth understandable by a general practitioner in your field of research, and demonstrate their relevance to core research problems in the field. This should be done discretely, and without being overly confident and critical of fellow researchers. Most program directors are willing to listen, but you should remember that their attention span may not be too long and also they have very busy schedules. The more succinct your arguments are, the more likely you will get your ideas across and leave a positive impression. In the end, if the meeting goes well, the program director will develop some affinity for your ideas, and will likely remember them during a panel evaluation and when the time comes to write a funding recommendation. If the meeting does not go well then you can move on and look for other programs for a better match for your proposal.

Following the proposal submission guidelines of funding agencies is equally critical for submitting competitive proposals. Every investigator who plans to submit a proposal to a funding agency ought to read the agency's proposal submission guidelines to avoid possible violations. In some cases, proposals may be returned without being reviewed if they do not meet a funding agency's guidelines. For

Table 6.1 Number of Proposals Returned to Investigators by NSF
without Review

Fiscal year	2003	2004	2005	2006	2007	2008
Number of proposals returned	276	236	176	134	117	124

Source: National Science Board, *Report to the National Science Board on the National Science Foundation's Merit Review Process, Fiscal Year 2008.*

example, NSF views intellectual merit and broader impact as the two most important criteria for all proposals. This is made clear with the following statement in NSF's *Grant Proposal Guide:** "Proposals that do not separately address both merit review criteria (intellectual merit and broader impact) within the one-page project summary will be returned without review."

Table 6.1 lists the number of proposals returned to investigators without review between 2003 and 2008. The NSF guide also includes other format and content-related specifications for project description, reporting results from prior NSF-funded research, qualifications of investigators, collaborative research, synergistic activities, page limitations, budget items, and current and pending support. It will be to the benefit of an investigator to go over these specifications during the submission of each proposal, as they tend to change over time.

6.3   Classification of proposals

After proposals are submitted to a funding agency in response to a program solicitation, they are classified into research areas either by program directors sifting through them or by an electronic assignment process that recognizes the program codes entered into the proposals. Despite the good intentions of all parties involved, many funding agencies handle thousands of proposals each year, and some proposals fall through the cracks without receiving an adequate review and evaluation due to a number of factors. One such factor is the inexperience or poor judgment of an investigator in identifying the right program for his or her proposal. The choice of program to which an experienced investigator submits a proposal is often reliable, but it is difficult to anticipate the mix of proposals that will be submitted during a solicitation cycle. A more dynamic process can help avoid the misclassification of proposals if proposals are sorted into panels based on keywords and other more refined identifiers. This is especially important for interdisciplinary proposals and broad agency solicitations that target a wide range of research interests.

---

* NSF Grant Proposal Guide, 2011, p. II-7.

Experienced program directors can identify lone proposals and act on them by soliciting mail reviews or asking another program director to handle them. However, in some cases, a proposal may end up on a panel with a bunch of proposals that have a very little subject overlap with it. This is typically bad news for that proposal, as the composition of the panel will likely not have enough expertise to evaluate it adequately. A less common but equally perilous situation arises when more than one program director shows a passing interest in a proposal. This often results in more than a typical number of reviews, with the proposal ending up on more than one panel. When this happens, the outcome is not likely to be favorable as the probability of two or more panels of experts having a strong endorsement for a proposal is pretty small. Rarely, a proposal may be of genuine interest to more than one program director rather than a desire to inflate his or her portfolio of proposals, and if this ever happens then that proposal will likely be funded.

These issues underscore the importance of proposal classification in funding agencies. The potential drawbacks of ad hoc proposal classification methods can be minimized if more comprehensive and structured clustering and panel assignment techniques are employed. The following two sections describe these techniques in detail.

6.3.1   Proposal clustering

Clustering arises naturally in proposal classification at funding agencies. A first cut for clustering is made when proposals are separated into main research divisions and possibly individual research programs by their subject classifications. For example, some of the proposals submitted to the Mathematical Sciences Division at NSF will likely be further classified into the Geometric Analysis program. If a single panel of reviewers can handle all the proposals received by this program, then no other work is required to classify them any further.* Otherwise, they would have to be clustered into a number of smaller sets of proposals based on panel size, reviewer availability, cost of holding panels, and subject overlap constraints. The first three constraints are often handled after proposals are separated into clusters using a subject classification method. The same ordering will be followed here and the first consideration will be the proposal clustering problem. Expert selection and panel assignment problems will be discussed in the next section.

---

* A typical program may receive between 100 and 200 proposals during each solicitation cycle. Theoretically, it is possible to assemble as large a panel as desired to deal with any number of proposals. However, in practice, several factors such as time and space restrictions as well as the coherence of panel deliberations place a limit on panel sizes. The size of proposal review panels this author attended rarely fell below 8 or exceeded 15 experts.

The subject overlap among a set of n proposals can be described by an $n \times n$ symmetric matrix $Y = [y_{ij}]_{n \times n}$, where

$$y_{ij} = y_{ji} = \begin{cases} 1 & if \ i = j \\ x_{ij}, \ 0 \le x_{ij} \le 1 & if \ i \ne j \end{cases} \tag{6.1}$$

The entry y_{ij} is called the affinity between proposals i and j, $1 \le i, j \le n$ and represents the extent of subject overlap between them. The affinities between proposals can be determined from the keywords they use or references they cite, or using some other content-matching process. Two proposals with no subject overlap may be assigned an affinity of 0 and those with complete subject overlap may be assigned an affinity of 1. If a ternary scale is used, then any two proposals may be assigned an affinity of 0, 0.5, or 1, indicating no overlap, some overlap, or complete overlap, respectively.

With these definitions in place, we can formalize proposal clustering as a process of dividing a set of proposals into disjoint sets under a number of affinity preserving metrics. It should be obvious that when proposals are separated into different clusters, some of the affinities will have to be removed. Equivalently, the remaining affinities will be preserved. The quality of this affinity preservation can be measured by a number of different metrics. One such metric focuses on limiting the total affinity needed to be removed to partition a set of proposals into a desired number of clusters of a desired size as described next.

Proposal clustering problem (minimum affinity cut): Given a set P of $n = uv$ proposals, with an affinity matrix $Y = [y_{ij}]_{n \times n}$, partition P into u subsets of proposals of size v, $P_1, P_2, \ldots, P_u$, where $n = uv$, and such that $\sum_{1 \le \alpha < \beta \le u} \sum_{i \in P_\alpha, j \in P_\beta} y_{ij}$ is the minimum over all such choices of u proposals. Thus, our goal is to minimize the sum of affinities across all pairs of subsets of proposals in the partition. Consequently, the total affinity that remains within all the clusters is maximized.

The proposal clustering problem is closely related to what is called the graph partitioning problem. Let $G = (V, E)$ be an undirected graph with a set of vertices, denoted V, and a set of edges, denoted E. Each edge in E is labeled by a proper fraction, that is, a number x, $0 < x \le 1$, which is called its weight. The vertices in V represent proposals, and the edges in E together with their weights represent the affinities between the pairs of proposals they connect together. The weight on an edge is interchangeably called the affinity of that edge. The combined affinity of a set of edges is the sum of affinities assigned to those edges. Missing edges represent affinities of 0 between proposals. The affinity of a vertex (proposal) is the sum of the affinities of all the edges connected to it. The combined affinity of a set of vertices (proposals) is the sum of the affinities of all those vertices. The affinity of a graph is the sum of the affinities of all its edges. With these interpretations of vertices and edges

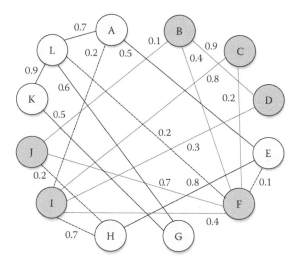

Figure 6.2 Clustering 12 proposals into two groups of 6 proposals.

in G, the proposal clustering problem is equivalent to partitioning G into u unconnected subgraphs of v vertices each by removing a subset of edges with as small a combined affinity as possible.*

Example 6.1

Consider the graph shown in Figure 6.2. The vertices, labeled A through L, denote 12 proposals with their affinities to other proposals indicated by the weights on the edges connecting them. Pairs of proposals (B, D), and (K, L) each have an affinity of 0.9, which is the highest among all pairs of proposals. On the other hand, pairs of proposals (B, J), and (E, F) each have an affinity of 0.1, which is the lowest among all pairs of proposals. The affinity of vertex A is 1.4, and the combined affinity of vertices A and B is 2.8. The goal of clustering in this example is to identify a set of edges whose removal would separate the graph into two parts of six vertices each and whose combined affinity is the smallest among all such sets of edges.

It is tempting to remove those edges with the smallest affinities first as they signify intangible overlaps between proposals, but this may not result in a partition with a cut of edges having a minimum combined affinity. For example, removing the edges with affinities 0.1, 0.1, 0.2, 0.2, 0.2, 0.2, and 0.3 would result in a combined affinity loss of 1.3, but doing so does not partition the graph into two subgraphs. On the other hand, the partition highlighted by the gray and dark gray vertices results in a removal of combined affinity of 1.4 and partitions the graph into two subgraphs of six vertices each. The dashed lines represent the five edges that are removed to generate this partition. It is seen that the removed edges include four of the seven lowest affinities but do not include one of the two edges with the minimum affinity, that is, edge (B,J). Furthermore, it includes an edge with a fairly large affinity, that is, edge (H,I).

---

* Aykanat, C., and Kayaaslan, E. 2011. Proposal and referee bi-clustering for panel formation. Technical Report, BU-CE-1104, Computer Engineering Department, Bilkent University.

A natural question to ask at this point is whether the partition shown in Figure 6.2 uses a cut of edges with the smallest possible combined affinity for any such partition. That this is indeed the case can be established using a lower bound on the number of edges needed to partition a graph into k parts.*

Remark 6.1 (Donath and Hoffman) Let $G = (V,E)$ be a graph with an $n \times n$ matrix[†], $Q_G = [q_{ij}]_{n \times n}$, where

$$q_{ij} = \begin{cases} -degree\ (i) & if\ i = j \\ 1 & if\ i \neq j\ and\ (i,j) \in E \\ 0 & if\ i \neq j\ and\ (i,j) \in E \end{cases} \tag{6.2}$$

Let e_c denote the number of edges whose removal from E partitions G into k pairwise disjoint subgraphs with $1 \leq i \leq k$ vertices. Then

$$e_c \geq -\frac{1}{2} \sum_{i=1}^{k} m_i\, \lambda_i\, (Q_G) \tag{6.3}$$

where, $\lambda_i(Q_G)$, $i = 1, 2 \ldots , k$ are the k largest eigenvalues of the Q_G.

For the graph in the previous example, $k = 2$, $m_1 = m_2 = 6$, and Q_G is a 12×12 matrix with entries

$$Q_G = \begin{bmatrix}
-3 & 0 & 0 & 0 & 1 & 0 & 0 & 0 & 1 & 0 & 0 & 1 \\
0 & -3 & 0 & 1 & 0 & 1 & 0 & 0 & 0 & 1 & 0 & 0 \\
0 & 0 & -2 & 0 & 0 & 1 & 0 & 0 & 1 & 0 & 0 & 0 \\
0 & 1 & 0 & -2 & 0 & 0 & 0 & 0 & 1 & 0 & 0 & 0 \\
1 & 0 & 0 & 0 & -3 & 1 & 0 & 1 & 0 & 0 & 0 & 0 \\
0 & 1 & 1 & 0 & 1 & -6 & 0 & 0 & 1 & 1 & 0 & 1 \\
0 & 0 & 0 & 0 & 0 & 0 & -2 & 0 & 0 & 0 & 1 & 1 \\
0 & 0 & 0 & 0 & 1 & 0 & 0 & -3 & 1 & 1 & 0 & 0 \\
1 & 0 & 1 & 1 & 0 & 1 & 0 & 1 & -5 & 0 & 0 & 0 \\
0 & 1 & 0 & 0 & 0 & 1 & 0 & 1 & 0 & -3 & 0 & 0 \\
0 & 0 & 0 & 0 & 0 & 0 & 1 & 0 & 0 & 0 & -2 & 1 \\
1 & 0 & 0 & 0 & 0 & 1 & 1 & 0 & 0 & 0 & 1 & -4
\end{bmatrix}$$

---

* Donath, W. E., and Hoffman, A. J. 1973. Lower bounds for partitioning of graphs. *IBM J. Res. Dev.* 9:420–425.

† The matrix Q_G is often referred to as the Laplacian of graph G in graph theory literature.

The two largest eigenvalues of Q_G can be shown to be -0.481 and -1.436, and substituting these values into Equation 6.3, we find

$$e_c \geq \left\lceil -\frac{1}{2}(m_1\lambda_1 + m_2\lambda_2) \right\rceil = \left\lceil -\frac{1}{2}(6\lambda_1 + 6\lambda_2) \right\rceil = \left\lceil -3(-0.481 - 1.436) \right\rceil = \left\lceil 4.917 \right\rceil = 5.$$

Thus, any partition of the graph shown in Figure 6.2 into two subgraphs of six vertices each requires the removal of at least five edges, matching the number of cut edges in the graph. This eliminates the possibility of obtaining such a partition with fewer than five edges in a cut regardless of the combined affinity of the removed edges. However, it is still possible to have a cut with more than five edges and smaller combined affinity, since the lower bound does not factor the weights of edges. The maximum number of edges in such a cut can be bounded by noting that the combined affinity of any eight or more edges cannot be less than the sum of the smallest eight affinities in the graph. This sum is equal to 1.7 for this graph and we already have a cut of edges with a combined affinity of 1.4. Therefore, the minimum partition of this graph into two subgraphs of six vertices each cannot include more than seven edges. Suppose that there is a cut of seven edges that produces such a partition. Then such a cut must necessarily include the edges with the smallest six affinities, that is, those edges with affinities 0.1, 0.1, 0.2, 0.2, 0.2, and 0.2 since excluding any one of these edges from the cut would require including two edges with a combined affinity of at least 0.7 (the two edges with the next two smallest affinities) that would then push the combined affinity beyond 1.4. This then only leaves three feasible sets of seven edges, one with a combined affinity of 1.3, and another two with a combined affinity of 1.4. The latter two do not improve the combined affinity of the cut edges in the partition in Figure 6.2. On the other hand, removing the first set of edges shown below does not result in a partition of the graph.

$$(A, I), (B, J), (C, F), (D, I), (E, F), (F, L), (H, J)$$

A similar analysis can be used to show that none of the cuts with six edges leads to a partition with a combined affinity of a cut of edges that is less than 1.4. Therefore, the cut of edges shown in Figure 6.2 produces a minimum partition as desired.

Minimizing the combined affinity of the set of edges in a cut is a good first step to divide a set of proposals into equal-size clusters. Doing so maximizes the combined affinity of the edges in the resulting clusters. However, this does not guarantee that all clusters of proposals would have the same combined affinity. This can be seen by summing the weights of edges in each subgraph in the aforementioned example.

The subgraph with dark gray vertices has a combined affinity of 3.8 whereas the subgraph with gray vertices has a combined affinity of 4.0. Even though the combined affinities of the two subgraphs are close together in this case, minimizing the sum of affinities over the removed edges may result in partitions with significantly uneven distributions of combined affinities.

Example 6.2

The graph in Figure 6.3 illustrates such a distribution, where all edge affinities are assumed to be 1. If we wish to minimize the combined affinity on the edges needed to partition the graph into two subgraphs of six vertices then we must use cut 1 in Figure 6.3a and remove the edges (C, G) and (D, L). This results in a combined affinity of 15 for the subgraph on the left and a combined affinity of 6 for the subgraph on the right. One consequence of this uneven distribution of combined affinities is that the proposals in the first subgraph have too much

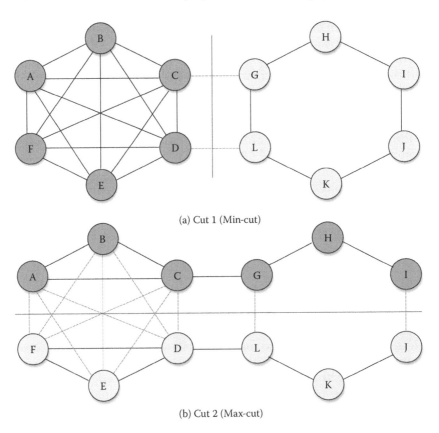

(a) Cut 1 (Min-cut)

(b) Cut 2 (Max-cut)

Figure 6.3 Clustering proposals with a minimum cut and maximum cut.

overlap, whereas those in the second subgraph have too little. Nonetheless, after the cut is made, the affinities within each subgraph are uniformly distributed over all proposals; each proposal in the first subgraph overlaps with all others, and each proposal in the second subgraph overlaps exactly with two proposals. This may be viewed as a good choice of dividing the 12 proposals evenly into two parts.

If an uneven distribution of affinities of proposals across clusters still remains a concern, different partitions may be used. For example, we may use cut 2 in the graph shown in Figure 6.3b to set the combined affinities of both subgraphs to 6. In a way, this cut maximizes the combined affinity of the removed edges to partition the twelve proposals into two subsets of six proposals, each with a combined affinity of 6. The uniformity of affinities across or within clusters of proposals may be further controlled by statistical metrics such as the mean and variance of the combined affinities of proposals.

To this end, we define the mean vertex (proposal) affinity of a set of n proposals with affinities, y_{ij}, $1 \le i \ne j \le n$, as

$$E_y = \frac{1}{n} \sum_{1 \le i \ne j \le n} y_{ij} \tag{6.4}$$

and its variance as

$$V_y = \frac{1}{n} \sum_{1 \le i \le n} \left(\left(\sum_{1 \le j \ne i \le n} y_{ij} \right) - E_y \right)^2 \tag{6.5}$$

These formulas can be used to determine how uniformly affinities distribute over a partition of proposals.

Example 6.3

Consider the graph in Figure 6.4. The mean vertex affinity of the 12 proposals and variance of their affinities before clustering are given by*

$$E_y = \frac{1}{12} \sum_{1 \le i \ne i \le 12} y_{ij} = 1. \tag{6.6}$$

$$V_y = \frac{1}{12} \sum_{1 \le i \le 12} \left(\left(\sum_{1 \le j \ne i \le 12} y_{ij} \right) - E_y \right)^2 = \frac{1}{12} \sum_{1 \le i \le 12} (1-1)^2 = 0. \tag{6.7}$$

* Note that the affinities of missing edges are 0.

We have three cuts of edges, each with a minimum combined affinity of 1. To determine which one distributes the affinities more uniformly over the 12 proposals, we compute their mean and variance in each case as shown in Table 6.2. It is seen that the third cut of edges gives the smallest variance of affinities of proposals around the mean. Even though this cut separates proposal *C* from all but one of the proposals with which it overlaps, it keeps it in the same cluster with the proposal with which it has the maximum overlap. This is also true for the second cut of edges. In contrast, the first cut of edges separates the most closely related pairs of proposals *C* and *G*, and *D* and *L* into different clusters.

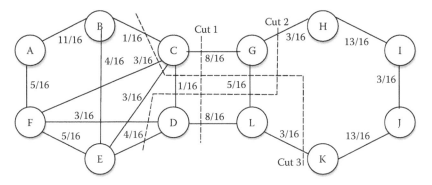

Figure 6.4 Clustering proposals using mean and variance metrics.

Table 6.2 Distribution of Combined Affinities for Three Cuts of Edges

Proposal	Affinity		
	Cut 1	Cut 2	Cut 3
A	1	1	1
B	1	1	15/16
C	8/16	15/16	8/16
D	8/16	8/16	15/16
E	1	12/16	13/16
F	1	1	13/16
G	8/16	8/16	11/16
H	1	13/16	13/16
I	1	1	1
J	1	1	1
K	1	1	13/16
L	8/16	11/16	8/16
Mean affinity	10	10.1875	9.8125
Variance	0.06	0.038	0.032
Standard deviation	0.246	0.196	0.177

In general, the graph partitioning problem is known to be NP-complete* but there exists several heuristics with polynomial time complexities, notably by Kernighan and Lin,[†] Fiduccia and Mattheyses,[‡] and others (see, for example, Karypis and Kumar[§]). A number of software packages are also available for partitioning graphs with weighted edges.[¶] Metis[**] was used to generate the partition in Figure 6.2. Metis provides two programs, kmetis and pmetis, to partition graphs into a desired number of subgraphs with nearly an equal number of vertices. The graph to be partitioned is entered using a text file and the generated output is also stored in a text file. The input and output files for the graph and its partition are shown in Figure 6.5.

The first line in the input file specifies the number of vertices, number of edges, and the option of having weights on edges, which is indicated by the "1" entry at the end. The lines that follow specify the edges and their weights beginning with those that are connected to the first vertex. For example, vertex A (denoted as vertex 1 in Metis) is connected to vertices 5, 9, and 12 (*E*, *I*, and *L*). The entries between these numbers are the weights 0.5, 0.2, and 0.7 multiplied by 10 as Metis only accepts integer weights. Other lines are filled in the same way to specify the edges for the remaining vertices and their weights. The output file represents the partition by marking the vertices with 0 and 1, where 0 represents the first subgraph and 1 represents the second one. The subgraphs can be constructed from this information since the edges are specified in the input file. Finally, the console output provides additional information about the partition. In this case, it lists the number of vertices, number of edges in the graph, and number of parts desired in the partition. It also lists the sum of the weights of the edges in the cut, and how balanced the two parts are with respect to the number of vertices they contain. In this example, the sum of the weights is $10 \times 1.4 = 14$. The 12 vertices are evenly divided between the two subgraphs. Therefore, the partition is 100% balanced as indicated by the value of 1.00 next to the word *balanced*. In general, the balance of a partition is given by n/km, where n is the number of the vertices in the

* Garey, M. R., and Johnson, D. S. 1979. *Computers and intractability: A guide to the theory of NP-completeness.* W.H. Freeman.
† Kernighan B. W., and Lin, S. 1970. An efficient heuristic procedure for partitioning graphs. *Bell Sys. Tech. J.* 49:291–307.
‡ Fiduccia, C. M., and Mattheyses, R. M. 1982. A linear time heuristic for improving network partitions, In *Proceedings of the 19th IEEE Design Automation Conference*, pp. 175–181.
§ Karypis, G., and Kumar, V. 1998. A fast and high quality multilevel scheme for partitioning irregular graphs. *SIAM J. Sci. Comp.* 20:359–392.
¶ Pellegrini, F. 1997. Graph partitioning methods and tools for scientific computing. *Parallel Comput.* 23:153–164.
** Karypis, G., and Kumar, V. 1998. Metis, version 4.0 [software]. Metis is a software package for partitioning unstructured graphs, partitioning meshes, and computing fill-reducing orderings of sparse matrices.

```
12 19 1

5  5  9  2  12

4  9  6  4  10

6  2  9

2  9  9

1  5  6  1  8

2  4  3  2  5  1  9  4  10  7  12

11  5  12

5  8  97  10

1  2  3  8  4  3  6  4  8

2  1  6  7  8

7  5  12

1  7  6  2  7  6  11  9
```

(a) Input

```
0  1  1  1  0  1  0  0  1  1  0  0  (Transposed)
```

(b) Output

```
METIS 4.0.1 Copyright 1998, Regents of the University of Minnesota

Graph Information --------------------------------------------------

  Name: propaffinity1Weighted.graph, #Vertices: 12, #Edges: 19, #Parts: 2

Recursive Partitioning... ------------------------------------------

  2-way Edge-Cut:      14, Balance:  1.00

Timing Information -------------------------------------------------

  I/O:                      0.000

  Partitioning:             0.000    (PMETIS time)

  Total:                    0.000

*****************************************************************
```

(c) Console output

Figure 6.5 Partitioning of the graph in Figure 6.2 using pmetis on Metis.

original graph, k is the number of subgraphs in the partition, and m is the number of vertices in the largest subgraph in the partition. When all parts have the same size then the partition is 100% balanced.

Metis and other partitioning software packages can partition affinity graphs of proposals easily as they handle graphs with many thousands of vertices.[*][†]

---

* Pellegrini, "Graph partitioning methods."
† Karypis and Kumar, "Metis."

6.4 Reviewer selection and panel assignment process

Funding agencies rely on three forms of review: (a) mail-only review, (b) panel-only review, (c) mail-and-panel review. Most proposals are evaluated using panel-only and mail-and-panel review methods as illustrated in the NSF statistics in Table 6.3.

Panels are assembled for a one-time evaluation of a set of proposals that have been bundled together by program directors using subject and program classification criteria that are set by a funding agency. They typically handle between 10 and 40 proposals. Once a set of proposals is clustered into panels, a program director seeks a set of reviewers to assemble a panel for each cluster of proposals. Reviewers are selected from a pool of researchers working at academic institutions, and government and private research laboratories.

The minimum number of reviewers for a panel of proposals, r, is often determined by the following formula:

$$r \geq \left\lceil \frac{n \times s}{k} \right\rceil \tag{6.8}$$

where n, s, and k denote the number of proposals, number of reviews required for each proposal, and number of proposals assigned to each reviewer, respectively. The values of s and k can be traded to control the number of reviewers for a given cluster of proposals. For example, if 20 proposals are placed in a cluster, each proposal is required to receive four reviews, and each reviewer is assigned to review 10 proposals then a panel of 8 reviewers is needed. A panel of 6 reviewers can be formed by decreasing s to 3 or increasing k to 15.

In general, assembling panels can be viewed as an assignment problem, where a set of reviewers is matched with a set of proposals in a cluster. If reviewers come from a fixed pool then the assignment problem is said to be constrained, and involves minimizing or maximizing a given objective function with predetermined capacities and specialties.

One such objective function is to maximize the number of pairs of proposals by a set of reviewers or cover all such pairs of proposals. A pair

Table 6.3 Funding and Review Statistics of NSF Proposals in 2008

	All Methods	Mail Only	Panel Only	Mail and Panel
Reviews	248,772	15,665	132,963	100,144
Proposals	42,983	3,662	24,966	14,355
Reviews/proposals	5.8	4.3	5.3	7

Source: National Science Board, *Report to the National Science Board on the National Science Foundation's Merit Review Process, Fiscal Year 2008*, p. 25.

of proposals is said to be covered by an assignment if they are assigned to the same reviewer. It is assumed that a reviewer who is assigned to review k proposals can compare all $\binom{k}{2} = k(k-1)/2$ pairs of proposals, and hence cover them. Thus, if proposals A, B, and C are assigned to a reviewer then all three pairs of proposals (A,B), (A,C), and (B,C) are covered by that reviewer. Being able to compare proposals in pairs provides an ordinal evaluation of a set of proposals without requiring all reviewers to compare all the proposals.[*,†,‡] In particular, if a panel consists of r reviewers to evaluate n proposals, and each reviewer is assigned to review k proposals, then the following inequality holds:

$$r \geq \left\lceil \frac{\binom{n}{2}}{\binom{k}{2}} \right\rceil = \left\lceil \frac{n(n-1)}{k(k-1)} \right\rceil \tag{6.9}$$

For example, if each reviewer is assigned to review 10 proposals from a set of 25 proposals then

$$r \geq \left\lceil \frac{25 \times 24}{10 \times 9} \right\rceil = 7 \tag{6.10}$$

In general this lower bound would be tighter than the one in Equation 6.8 if $s < (n-1)/(k-1)$. Even so, this lower bound is not always attainable. For example, if $n = 4$, and $k = 3$, then $r \geq 2$ but using two reviewers is not sufficient to cover all six pairs of proposals in this case. To see this, consider a set of proposals $\{p_1, p_2, p_3, p_4\}$ and without loss of generality, suppose that the first reviewer is assigned the first three proposals, p_1, p_2, and p_3. To include the last proposal, the second reviewer must review proposal p_4, and two more proposals that overlap with those reviewed by the first one. Thus, the possible choices of proposals for the second reviewer are $\{p_1, p_2, p_4\}$, $\{p_1, p_3, p_4\}$, $\{p_2, p_3, p_4\}$. However, combining any of these three assignments with the assignment of p_1, p_2, and p_3 to the first reviewer leaves out one pair of proposals (respectively, (p_3, p_4), (p_2, p_4), (p_1, p_4)). As we shall see later, this lower bound can be improved by other combinatorial arguments.

---

* Cook, W. D., Golany, B., Kress, M., Pen, M., and Raviv, T. 2005. Optimal allocation of proposals to reviewers to facilitate effective ranking. *Manag. Sci.* 51:655–661.
† Oruc, A.Y., and Atmaca, A. 2009. Asymptotically optimal assignments in ordinal evaluations of proposals. arXiv:0908.3233v1[cs.DM].
‡ Atmaca, A., and Oruc, A. Y. 2010. Ordinal evaluation and assignment problems. *IEEE Int. Conf. Sys. Man Cyber*, 3284–3289. DOI: 10.1109/ICSMC.2010.5642346.

6.4.1   Panel assignments with capacity and specialty constraints

In some cases, program directors may prefer to work with a fixed pool of reviewers whose research specialties are also fixed. In this case, the panel assignment problem can be stated in terms of (1) an $r \times n$ binary matrix, $X = [x_{ij}]_{r \times n}$, where $x_{ij} = 1$ if and only if reviewer i can be assigned to review proposal j, $1 \le i \le r$, $1 \le j \le n$, and (2) an r-element vector, $K = [k_i]$, where k_i denotes the maximum number of proposals that can be assigned to reviewer i, $1 \le i \le r$. X is called the reviewer–proposal pairing matrix. K is called the reviewer capacity vector, and k_i is called the capacity of reviewer i. The entries in X match the specialties of reviewers with pro-posals whereas the entries in K specify capacity restrictions of reviewers. The capacities of reviewers may clearly affect the maximum number of pairs of proposals that can be covered by a given set of reviewers. The fol-lowing example illustrates that the distribution of their specialties among the proposals is also critical.

Example 6.4

Let $n = 8$, $r = 4$. Rearranging the terms in Equation 6.9, we have $k(k - 1) \ge n(n - 1)/r = 56/4 = 14$ or $k \ge 5$. Therefore, each reviewer should have a capacity of at least 5 to cover all pairs of proposals. To meet this minimum capacity bound for each reviewer, suppose that the 8 proposals are assigned to a panel of 4 review-ers according to the following reviewer–proposal pairing matrix in which each reviewer is assigned exactly 5 proposals:

$$
X = \begin{array}{c} \\ r_1 \\ r_2 \\ r_3 \\ r_4 \end{array}
\begin{array}{c} \begin{matrix} p_1 & p_2 & p_3 & p_4 & p_5 & p_6 & p_7 & p_8 \end{matrix} \\
\begin{bmatrix} 1 & 1 & 0 & 0 & 1 & 1 & 1 & 0 \\ 0 & 1 & 1 & 1 & 0 & 1 & 0 & 1 \\ 1 & 0 & 0 & 1 & 1 & 0 & 1 & 1 \\ 0 & 1 & 1 & 0 & 1 & 1 & 1 & 0 \end{bmatrix} \end{array}
\qquad (6.11)
$$

As defined earlier, the "1" entries in the matrix specify the proposals each reviewer is qualified to evaluate. For example, reviewer r_1 is qualified to evalu-ate proposals p_1, p_2, p_5, p_6, and p_7. Scanning the matrix row by row and from top to bottom shows that each pair of proposals, with the exception of (p_1, p_3), is covered by at least one referee. However, this does not mean that we need to increase the capacities of the reviewers, as it is possible to cover all pairs of proposals by switching r_1 from proposal p_5 to proposal p_3. Thus the distribution of the specialties of reviewers over proposals is as critical as the number and capacities of reviewers to cover all pairs of proposals.

In general, the capacities of reviewers may not be as large as the number of proposals they are qualified to review and this leads to a con-strained optimization problem as defined next.

Constrained panel assignment problem: Given a set of n proposals and a set of r reviewers, together with an $r \times n$ reviewer–proposal pairing matrix, $X = [x_{ij}]_{r \times n}$, and r-element reviewer capacity vector, $K = (k_1, k_2, ..., k_r)$, find an assignment that maximizes the number of pairs of n proposals covered by the r reviewers without exceeding the reviewer capacity limits specified in K.

Thus, if X_i denotes the subset of proposals that reviewer r_i is qualified to evaluate, then the goal of the constrained panel assignment is to select k_i proposals from X_i in such a way that the number of pairs of proposals covered between all r reviewers is maximized. The problem becomes trivial if we set $k_i = |X_i|$, $1 \le i \le r$, where $|X_i|$ denotes the number of proposals in X_i, as this would amount to computing all pairs of proposals generated by all the reviewers and taking their union. In other cases, and in particular as k_i gets close to $|X_i|/2$, the size of the search space of assignments, N_a, can become very large as the following formula shows:

$$N_a = \prod_{i=1}^{r} \binom{|X_i|}{k_i} \tag{6.12}$$

Example 6.5

In the previous example, we had assumed that $r = 4$, $|X_i| = k_i = 5$. If we change k_i to 4 then we have

$$N_a = \prod_{i=1}^{4} \binom{5}{4} = 5^4 = 725 \tag{6.13}$$

possible assignments. These choices of assignments are generated by the fact that each reviewer can potentially be assigned any 4 of the 5 proposals independently of others. If we increase the size of X_i to 6, effectively replacing one of the 0 entries by 1 in each row of X in Equation 6.11, then N_a becomes much larger:

$$N_a = \prod_{i=1}^{4} \binom{6}{4} = 15^4 = 50625 \tag{6.14}$$

Thus, the size of the search space of all possible assignments not only grows exponentially with the number of referees, but it can also grow very rapidly as the ratio $k_i/|X_i|$ approaches $1/2$. More precisely if we let $|X_i| = q$ and $k_i = k$, $1 \le i \le r$, then by Stirling's approximation,* we have

---

* Gallagher, R. G. 1968. *Information theory and reliable communication.* John Wiles & Sons, p. 530.

$$N_a = \prod_{i=1}^{4} \binom{q}{k} = \binom{q}{k}^r \geq \left\{ \left(\frac{q}{8k(q-k)} \right)^{1/2} e^{qH(\frac{k}{q})} \right\}^r \qquad (6.15)$$

where $H(\alpha) = -\alpha \ln \alpha \, (1-\alpha) \ln (1-\alpha)$, $0 < \alpha < 1$ is the entropy function. If we let, $q = 2k$ then

$$N_a \geq \left\{ \left(\frac{1}{2q} \right)^{\frac{1}{2}} e^q \right\}^r = e^{\left(q - \frac{\ln 2q}{2} \right) r} \qquad (6.16)$$

Thus, for $q = 2k = 10$ ($k = 5$), and $r = 10$, the search space of assignments exceeds $e^{88.5} > 10^{38}$.

The constrained panel assignment problem can be solved using branch and bound or integer programming techniques. In what follows, we describe an integer programming formulation that was suggested by Alper Yıldırım.* As before, we will use X_i and k_i to denote the subset of proposals that can potentially be assigned to reviewer i, and capacity of reviewer i, $1 \leq i \leq r$.

Constrained panel assignment problem—Integer programming (IP) formulation:

1. Decision variables:

$$\alpha_{ip} = \begin{cases} 1, \text{ if reviewer } i \text{ is assigned to proposal } p \\ 0, \text{ otherwise} \end{cases} \quad i = 1, 2, \cdots, r; p \in X_i$$

$$\beta_{ipq} = \begin{cases} 1, \text{ if reviewer } i \text{ is assigned to proposals } p \text{ and } q \\ 0, \text{ otherwise} \end{cases} \quad i = 1, 2, \cdots, r; p \in X_i; p \neq q$$

$$\gamma_{pq} = \begin{cases} 1, \text{ if the proposal pair } (p, q) \text{ is evaluated by at least one referee} \\ 0, \text{ otherwise} \end{cases} \quad p \neq q$$

2. IP formulation:
   2.1. Maximize subject to

$$\sum_{p, q \in P} \gamma_{pq}$$

   2.2. $\displaystyle\sum_{p \in X_i} \alpha_{ip} \leq k_i, \ i = 1, 2, \cdots, r$

   2.3. $\beta_{ipq} \leq \alpha_{ip}, \ i = 1, 2, \cdots, r; \ p, q \in X_i, \ p \neq q$

* Alper Yıldırım, personal communication, April 6, 2011.

Table 6.4 Assignment of $n = 2k$ Proposals to Six Reviewers
Each with Capacity k

Proposals	$p_{1},...,p_{k/2}$	$p_{k/2+1}\,...,p_k$	$p_{k+1}\,...,p_{k+\lceil k/2\rceil}$	$p_{k+\lceil k/2\rceil+1}\,...,p_{2k}$
Reviewer r_1	k proposals			
Reviewer r_2		k proposals		
Reviewer r_3	$\lceil k/2\rceil$		$\lceil k/2\rceil$	
Reviewer r_4	$\lceil k/2\rceil$			$\lfloor k/2\rfloor$
Reviewer r_5		$\lfloor k/2\rfloor$	$\lceil k/2\rceil$	
Reviewer r_6		$\lfloor k/2\rfloor$		$\lfloor k/2\rfloor$

2.4. $\beta_{ipq} \leq \alpha_{ip},\ i = 1, 2, \cdots, r;\ p, q \in X_i,\ p \neq q$

2.5. $\gamma_{pq} \leq \displaystyle\sum_{p,\,q\in X_i,\,1\leq i\leq r} \beta_{ipq},\ p \neq q$

2.6. $\alpha_{ip} \in \{0, 1\},\ i = 1, 2. \ldots, r$

2.7. $\beta_{ipq} \in \{0, 1\},\ p, q \in X_i,\ p \neq q;\ i = 1, 2. \ldots, r$

2.8. $\gamma_{pq} \in \{0, 1\},\ p, q \in X_i,\ p \neq q;\ i = 1, 2. \ldots, r$

The objective function 2.1 maximizes the number of different pairs of proposals evaluated by at least one reviewer. Constraint 2.2 ensures that the capacities of reviewers are not exceeded. Constraints 2.3 and 2.4 guarantee that the decision variables α_{ip} and β_{ipq} are consistent. Constraint 2.5 makes sure that a pair of proposals (p,q) is counted in an assignment if and only if p and q are assigned to at least one common reviewer. Finally, constraints 2.6, 2.7, and 2.8 declare the ranges of each decision variable. The solution of this IP formulation produces the largest set of pairs of proposals each of which is assigned to at least one reviewer.

6.4.2   Panel assignments without specialty constraints

Panel assignments without specialty constraints may be used when there is a large pool of reviewers with a multitude of specialties or when proposals can be evaluated without any specialty requirements.[*][†] In this case, the goal is to cover all pairs of proposals with as few reviewers as possible rather than maximizing the number of pairs of proposals for a given reviewer–proposal pairing matrix as formalized next:

Unconstrained panel assignment problem: Given a set of n proposals, find an assignment that covers all $\binom{n}{2}$ pairs of n proposals using as few reviewers as possible, and each with a capacity of k.

[*] Oruc and Atmaca, *Asymptotically optimal assignments.*
[†] Atmaca and Oruc, *Ordinal evaluation.*

	p_1	p_2	p_3	p_4	p_5	p_6
r_1	1	1	1	0	0	0
r_2	0	0	0	1	1	1
r_3	1	1	0	1	1	0
r_4	1	1	0	0	0	1
r_5	0	0	1	1	1	0
r_6	0	0	1	0	0	1

(a) $n = 6, r = 6, k = 3$

	p_1	p_2	p_3	p_4	p_5	p_6	p_7	p_8
r_1	1	1	1	1	0	0	0	0
r_2	0	0	0	0	1	1	1	1
r_3	1	1	0	0	1	1	0	0
r_4	1	1	0	0	0	0	1	1
r_5	0	0	1	1	1	1	0	0
r_6	0	0	1	1	0	0	1	1

(b) $n = 8, r = 6, k = 4$

Figure 6.6 Assignments with complete covering of all pairs of proposals.

The unconstrained panel assignment problem is complicated by the fact that the optimality of an assignment depends on the availability of a tight lower bound for arbitrary k. Here, we present one such assignment* by first establishing a tight lower bound on the number reviewers when $n = 2k$.

Remark 6.2   For all $n = 2k > 4$, if each reviewer is assigned k proposals, at least six reviewers are needed to cover all pairs of n proposals.

Proof: See Oruc and Atmaca.[†,‡]

We now provide an explicit assignment that matches the lower bound of six reviewers.

Remark 6.3   For any even integer $n = 2k > 4$, if four reviewers are assigned k proposals each, one reviewer is assigned $2\lceil k/2 \rceil$ proposals and one reviewer is assigned $2\lfloor k/2 \rfloor$ proposals as shown in Table 6.4,[§] then six reviewers are sufficient to cover all pairs of n proposals.

Proof: See Oruc and Atmaca.[¶,**]

Example 6.6

The assignments shown in Figure 6.6 illustrate this assignment for six and eight proposals. In both cases, six reviewers are used as the number of reviewers is fixed to six for any even number of proposals.

This result can be extended to odd values of n as well as to other reviewer capacities as described in the next four remarks.[††,‡‡]

* Oruc and Atmaca, *Asymptotically optimal assignments.*
† Oruc and Atmaca, *Asymptotically optimal assignments.*
‡ Atmaca and Oruc, *Ordinal evaluation.*
§ Note that $\lceil k/2 \rceil + \lfloor k/2 \rfloor = k$.
¶ Oruc and Atmaca, *Asymptotically optimal assignments.*
** tmaca and Oruc, *Ordinal evaluation.*
†† Oruc and Atmaca, *Asymptotically optimal assignments.*
‡‡ Atmaca and Oruc, *Ordinal evaluation.*

Remark 6.4   For all $n = 2k + 1 > 5$, suppose that three reviewers are assigned $k + 1$ proposals each, and other reviewers are assigned k proposals each. Then at least six reviewers are both necessary and sufficient to cover all pairs of n proposals.

Remark 6.5   For all $n = 3k > 12$, if each reviewer is assigned k proposals, where k is divisible by 3, then at least 11 reviewers are necessary and 12 reviewers are sufficient to cover all pairs of n proposals.

Remark 6.6   For all $n = 4k > 16$, if each reviewer is assigned k proposals, where k is divisible by 4, at least 18 reviewers are necessary and 20 reviewers are sufficient to cover all pairs of n proposals.

Remark 6.7   Let n and k be positive integers, where k is even and divides n. It is sufficient to have $n(2n - k)/k^2$ reviewers, each with capacity k to cover all $n(n - 1)/2$ pairs of n proposals.

The assignments described in these remarks can be found in Oruc and Atmaca.* The last assignment works for any $k \le n/2$ and the number of reviewers it employs remains within a factor of two of the lower bound given in Equation 6.9.

6.5   *Proposal review and evaluation process*

The proposal evaluation process begins in earnest once proposals are clustered into panels, and reviewers and panels are determined. In many funding agencies, proposals are sent out to reviewers before they are paneled for final evaluation. Each reviewer is assigned a subset of all proposals in a cluster, and asked to evaluate them using a quantitative scale that may range from 0 to 3 or 5, or a qualitative scale that may include ratings such as poor, fair, good, very good, and excellent. Some reviewers assign intermediate ratings such as poor to fair, fair to good, good to very good, and very good to excellent.

Reviews are generated by mail reviewers or panelists prior to a panel meeting, and submitted electronically to a funding agency's proposal processing system as in the NSF's online fastlane system.† Subsequently, panels are convened at funding agencies to collectively evaluate the proposals that have been assigned to them. A lead reviewer, called a scribe, is assigned to each proposal to keep track of the deliberations that pertain to that proposal. Reviewers are generally allowed to change their ratings and comments during panel meetings. In the end, the panel is asked to rank all the proposals into a number of categories such as (a) highly

---

* Oruc and Atmaca, Asymptotically optimal assignments.
† Fastlane, www.fastlane.nsf.gov.

competitive, (b) competitive, and (c) noncompetitive categories or alternatively into (a) fundable and (b) nonfundable categories. In the first scheme, program directors and panel moderators may ask panels to further rank proposals in the competitive category, and in some cases, in the highly competitive category as well to ensure that proposals are prioritized for remaining funds after the proposals with the highest ratings have been funded.

Proposals in the highly competitive category have the highest probability of funding, and are more frequently funded than not. Those in the competitive category have a good probability of funding but much depends on the number of highly competitive proposals and the availability of funds. Noncompetitive proposals are rarely funded even though exceptions exist. The NSF statistics shown in Table 6.5 confirm these remarks. More than 70% of proposals in the excellent category were funded, and about 43% of those in the very good-to-excellent category were funded. Only about 15% of the proposals in the good-to-very good category were funded. The odds of winning an award were much smaller for the proposals in the fair-to-good and poor-to-fair categories. Overall, NSF funding rates have been fluctuating around 25%.

It is critical for researchers to know at what rate a funding agency awards grants. Ideally, every researcher would like to have all of his or her proposals funded but no funding agency can meet such an expectation. A more reasonable statistic is to get every other proposal funded or even one out of three proposals funded. When the rate drops below 30%, as is the case at NSF, some researchers may not bother to keep trying beyond three submissions. Funding agencies generally realize that low funding rates may drive researchers to other funding agencies or stop submitting proposals, even though they are rarely short supplied when it comes to proposals. Still, it is important for funding agencies to fund as many qualified proposals as possible. The statistics shown in Table 6.5 reveal that 28% of excellent proposals were declined at NSF in 2008. This is a significant declination rate for excellent proposals. Excellent proposals clearly deserve funding, but if this is not realistic because of lack of funds, it may be better to calibrate the review process so that fewer proposals are given excellent ratings.

6.6   Ranking proposals

One of the most contentious tasks that panels deal with is to rank the highly competitive and competitive proposals. In a nutshell, this is an ordinal ranking process where highly competitive and competitive proposals are compared and ranked with adjectives such as "better than," "more deserving," "higher potential for impact," and "more likely to succeed." Program directors often ask their panels to rank the proposals in order

Table 6.5 NSF Award Statistics and Funding Rates

Award Statistics of NSF Proposals, Fiscal Year 2008[a]

Outcome	Poor to Fair	Fair to Good	Good to Very Good	Very Good to Excellent	Excellent
Declines	1253	8981	2272	3398	1345
Awards	2	76	15814	5393	3445

Funding Rates at NSF[b]

Year	2002	2003	2004	2005	2006	2007	2008
Proposals	35,165	40,075	43,851	41,722	42,352	44,577	44,428
Awards	10,406	10,844	10,380	9,757	10,425	11,463	11,149
Funding Rate	30%	27%	24%	23%	25%	26%	25%

[a] National Science Board, *Report to the National Science Board on the National Science Foundation's Merit Review Process, Fiscal Year 2008*, p. 28.

[b] National Science Board, *Report to the National Science Board on the National Science Foundation's Merit Review Process, Fiscal Year 2008*, p. 36.

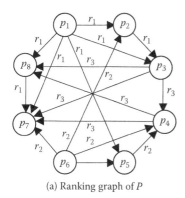
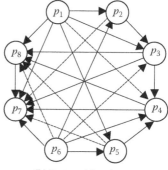

(a) Ranking graph of *P*   (b) Its transitive closure

Figure 6.7 A ranking graph of a set of proposals and its transitive closure.

to prioritize their recommendations. This is often done by placing the proposals up and down a list until panelists reach a consensus. A more methodical approach would be to ask panelists to independently rank all competitive and highly competitive proposals and then form a directed graph to complete the ranking process as described next.

Let $P = \{p_1, p_2, ..., p_n\}$ be a set of proposals, and $R = \{r_1, r_2, ..., r_m\}$ be a set of reviewers with a set of rankings, $\preceq_R = \{\preceq_{r_1}, \preceq_{r_2}, ..., \preceq_{r_m}\}$. For each ranking, $\preceq_{r_i}$, define a directed graph $G_i = (V, E_i)$, where $V = P$, and $E_i = \{(p_x, p_y) \in \preceq_{r_i}$ and $x \neq y, 1 \leq i \leq m\}$. Let $G_R = (V, E)$ be the union of graphs $G_i, 1 \leq i \leq m$ that is, let $V = P$ and $E = E_1 \cup E_2 \cup ... \cup E_m$, and call it the ranking graph of *P*.

The ranking graph represents the ordinal comparisons of a set of proposals by a set of reviewers. The ranking provided by each reviewer establishes a relation between all pairs of proposals to which he or she is assigned. In order to combine the rankings of all reviewers, we can use the transitive closure of a ranking graph. The transitive closure of a ranking graph is obtained by inserting a new directed edge between any two vertices p_i and p_j whenever there is a sequence of relations $p_i \preceq p_{x_1}, \preceq ... p_{x_q} \preceq p_j$.*

Example 6.7

Let $P = \{p_1, p_2, p_3, p_4, p_5, p_6, p_7, p_8\}$ and $R = \{r_1, r_2, r_3\}$ with the rankings

$\preceq_{r_1} = \{(p_1 \preceq p_2), (p_2 \preceq p_3), (p_1 \preceq p_3), (p_1 \preceq p_5), (p_1 \preceq p_8), (p_8 \preceq p_7), (p_1 \preceq p_7)\}$

$\preceq_{r_2} = \{(p_6 \preceq p_2), (p_6 \preceq p_5), (p_5 \preceq p_4), (p_6 \preceq p_4), (p_6 \preceq p_7)\}$

$\preceq_{r_3} = \{(p_3 \preceq p_4), (p_4 \preceq p_8), (p_3 \preceq p_8), (p_3 \preceq p_7), (p_4 \preceq p_7)\}$

The ranking graph of *P* is shown in Figure 6.7a, where the labels next to the directed edges indicate the reviewers that provide the comparisons of the proposals (vertices) to which they are assigned. For example, the label over the

---

* Here, $p_x \preceq p_y$ denotes that proposal p_y is preferred over proposal p_x.

directed edge between p_1 and p_2 indicates that reviewer r_1 prefers proposal p_2 to p_1. The graph in Figure 6.7b depicts the transitive closure of the ranking graph of P, where the dashed lines are added to indicate the transitive relations between the proposals.

The transitive closure provides a convenient way to check if a partial ordering (ranking) of a set of proposals is feasible. To do this, all we need is to make sure that the transitive closure graph does not have any cycles of two vertices. For n proposals, this can be done by checking possible cycles between all pairs of vertices in $O(n^2)$ time. Alternatively, we can check if the ranking graph or its transitive closure has a topological traversal. It is well known that a directed graph does not have any cycles if and only if it exhibits a topological traversal, that is, an ordering of its vertices that respects the successor (or predecessor) relations between them. Thus, we can use any topological traversal algorithm to determine if a ranking graph of a set of proposals contains a cycle. Most of these algorithms work by repetitively removing vertices with no incoming (or outgoing) edges until all the vertices are sorted. The time complexity of a typical topological traversal algorithm is $O(|V| + |E|)$. In the case of a set of n proposals, this translates to $O(n^2)$ time complexity as there exist $O(n^2)$ edges between n vertices.

For the ranking graph shown in Figure 6.7 one such topological traversal is $p_1, p_6, p_2, p_3, p_5, p_4, p_8, p_7$ and represents a possible ranking of the eight proposals from worst to best. Another topological traversal is $p_1, p_6, p_2, p_5, p_3, p_4, p_8, p_7$. In fact, there are four topological traversals in all, and they can be specified by the following logic expression, where the plus (+) sign represents the logical or function:

$$(p_1 + p_6), (p_2, p_3 + p_5), p_4, p_8, p_7.$$

Combining all four traversals result in the partial ordering given in Figure 6.8. Thus, p_7 is the most competitive, and p_1 and p_6 are the least competitive proposals.

In many cases, the transitive closure of a ranking graph of a set of proposals will likely include cycles of two vertices. Such cycles would indicate the presence of direct or transitive disagreements between reviewers. It is possible to obtain a partial ordering from such transitive closures of ranking graphs by removing some of the edges. This effectively amounts to discounting the comparisons of some proposals, and it is therefore desirable to minimize the number of removed edges. It can be shown that the number of removed edges cannot exceed $b - n + 1$, where b is the number of edges and n is the number of vertices in a ranking graph.

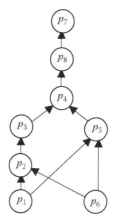

Figure 6.8 Ranking of proposals based on the set ranking shown in Figure 6.7.

6.7   Grant management and measuring effectiveness of funding

Different funding agencies have different grant management and project monitoring styles. Funding agencies that support basic research projects provide broad responsibilities to investigators in conducting their research projects. They probe the progress of research projects they fund by annual reports. They require a final report to assess if projects have been able to accomplish their research goals as well. Some funding agencies also require investigators to describe the research results that were obtained during a prior project they supported. All these methods help funding agencies make more effective funding decisions.

Funding agencies driven by a need to accomplish specific research goals rely more heavily on project monitoring and site visits. Project monitoring may motivate some investigators to get their research projects completed in time. Nonetheless, project monitoring should be carried out with care to avoid micromanaging research projects. This may potentially make some investigators think that they are working for a funding agency and they have a manager to report to every once in a while. This could ultimately impact the creativity of investigators and lead to mediocre research projects with marginal results. It is essential for researchers to have an independent track of thought and workspace to generate original results without being distracted by constant probing of funding agencies.

Measuring effectiveness remains a serious problem at funding agencies as research is an open-ended process, and the impact of a research project can only be measured in time. Some funding agencies conduct internal evaluations of their research programs annually and revise their solicitations to follow emerging research areas. They also ask committees of visitors (COVs) to have more comprehensive evaluations of their research programs and divisions. Such committees provide a list of recommendations for improving the funding effectiveness and other shortcomings of research programs.

Ultimately what matters for any funding agency is how much impact their funding makes in the research fields they invest. Although it is true that not all research projects can make a significant impact in their fields, funding agencies need to develop quantitative measures, much like universities do, to assess the success of their research programs. These measures may rely on similar metrics such as the number of graduate students that successfully complete their degree programs, journal articles published, conference papers presented, patents filed by investigators, citations received, and utility of research results in the development of new technologies and products, all accomplished with the help of research funds they provide. Clearly, these metrics cannot be compiled for proposals that are under evaluation, but funding agencies should have a track record of the investigators they have funded. These statistics can thus be gathered for funded projects by individual research programs, divisions, and even directorates. They can then be used to see how different programs and divisions compare over a period of time that may be adjusted to factor in the delay between the initiations of funding and anticipated times of research results of funded projects and their impact. The performance metrics and models that were described in Chapter 5 can easily be extended to carry out such statistical analyses of funding effectiveness of research programs at funding agencies. For example, the funding effectiveness in the three divisions (Computing and Communication Foundations, Computer and Network Systems, Information & Intelligent Systems) in the Computer and Information Science and Engineering Directorate at NSF can be compared using the aforementioned performance metrics. Of course, carrying out this process requires real data and is best left as an exercise for program directors and other funding policy experts that work in funding agencies.

6.8 Role of program directors in proposal evaluation and funding decisions

Program directors play a very critical part in funding decisions. They solicit reviews, assemble panels, and, write a final recommendation for every proposal submitted to their programs.

At NSF, a sizeable number of program directors are researchers that come to work for the federal government under the Intergovernmental Personnel Act (IPA). Of the 520 program directors in 2008, 212 were hired under the IPA, 215 were permanent, 47 were visiting scientists and educators, and the remaining 46 were temporary appointments.* Whether they are IPA contracted researchers or permanent employees, program directors need to be proactive as much as possible to keep up with the fast moving and emerging research topics. The job of a program director is not only central to the success of a funding agency, but it also has a significant impact on the professional development of researchers. Program directors need to be careful not to overload a small group of researchers at the expense of starving the broad base of researchers. This can drive people out of a field while rewarding a small clique of researchers serving one another. Proposal submission forms include a section where investigators list their current and pending funds. An experienced program would pay considerable attention to the funding level of an investigator before committing a funding decision. At the same time, funds should never be distributed uniformly as this will promote mediocrity, and the most prolific researchers will not have sufficient funds to accomplish their research goals. Knowing how to balance these conditions is a must for an effective program director.

The following guidelines may help program directors in improving the effectiveness of their decisions.

- Always state the conflict of interest rules to each panel before panelists begin to evaluate proposals. Most funding agencies require that panelists sign a conflict of interest form, but this is not enough of a deterrent for some reviewers especially when funding gets highly competitive.
- Gently warn panelists when they make subjective remarks about investigators. Typical examples are:
  - "The PI does not know what he is talking about."
  - "He is a second rate researcher."
  - "She is not qualified to work in the field."

---

* National Science Board, *Report to the National Science Board on the National Science Foundation's merit review process, fiscal year 2008*, p. 30.

- "I read one of his papers and it was really bad."
- "He should prove himself before submitting a proposal."

- Solicit sufficient number of reviews for each proposal to reach a strong consensus, and make sure that all proposals receive approximately the same number of reviews. An uneven distribution of reviews over proposals may lead to unfair and erroneous decisions.
- Do not allow a panelist to dominate the discussion on any given proposal.
- Read between the lines when incorporating the reviewer reports into a final recommendation. Discount sentences such as
    - "I feel that there should be an easier way to solve this problem."
    - "The research plan seems vague."
- "The literature survey does not cite my work or the work of someone I know."

Funding agencies often provide guidelines for program directors as well. The role of a program director at NSF is defined as follows:*

- Selects reviewers and panel members, based on program officer's knowledge, references listed in the proposal, individuals cited in recent publications or relevant journals, presentations at professional meetings, reviewer recommendations, bibliographic and citation databases, and proposal author's suggestions.
- Checks for conflicts of interest. In addition to checking proposals and selecting reviewers with no apparent potential conflicts, NSF staff provides reviewers guidance and instruct them how to identify and declare potential conflicts of interest. All NSF program officers receive annual conflict of interest training.
- Synthesizes the comments of the reviewers and panel (if reviewed by a panel), as provided in the individual reviewer analyses and panel summaries.
- Makes a recommendation to award or decline the proposal, taking into account external reviews, panel discussion, and other factors such as portfolio balance and amount of funding available.

In making their recommendations, NSF program directors are accorded sufficient autonomy but they are expected to make their recommendations using the following criteria[†]:

- Support for potentially transformative advances in a field

---

* National Science Board, p. 21.
† National Science Board, p. 23.

- Novel approaches to significant research questions
- Capacity building in a new and promising research area
- Potential impact on the development of human resources and infra-structure
- NSF core strategies, such as (1) the integration of research and education and (2) broadening participation
- Achievement of special program objectives and initiatives
- Other available funding sources
- Geographic distribution

The success of any funding process hinges critically on how well it matches the mission of a funding agency. This can only happen if program directors combine their professional experiences with the guidelines set forth by the funding agencies at which they work.

6.9   Five mistakes funding agencies should avoid to improve their impact

Funding agencies make an all-out effort to make sure that their policies and practices achieve their funding goals. Still, some do better than others. The following describes some of the most common issues that funding agencies need to address to improve the effectiveness and impact of their funding goals.

6.9.1   Mistake 1: Failing to predict when to start and stop funding a field of research

One of the major dilemmas funding agencies face is that high impact research projects do not come around frequently. Since research is an open-ended process, predicting which research proposal(s) may end up having a strong impact is difficult. This makes funding agencies reluctant to fund proposals that are too far out from the mainstream of ongoing research projects. This then encourages researchers to play it safe and submit proposals that focus on incremental research problems. With funds flowing into their research projects, many researchers get too comfortable and become reluctant to work on new research problems and initiate new ideas. In the end, funding agencies find themselves overly invested in incremental and mediocre research projects. To break this stagnant cycle, funding agencies can keep track of milestone discoveries and train their program directors to guard against incremental research proposals and be bold to fund high-risk and high-impact research proposals more frequently. They can also be proactive in identifying new frontiers of research by organizing workshops and meetings with broad participation from researchers in fundamental fields of science and engineering. Failing to do so would further marginalize contributions of researchers and result in wasting precious funding resources.

6.9.2   Mistake 2: Issues with peer review

Peer review is a popular evaluation system and used by most funding agencies in one form or another. With thousands of proposals handled by funding agencies each year, peer review is perhaps the only system of evaluation that makes sense. It allows the broadest participation of researchers in the process of deciding who should be funded and as such it represents the wisdom of all researchers in a given field. At the same time, it is a system that is potentially open to misuse and potential conflicts of interest. It is important for funding agencies to safeguard the peer review process using strict conflict of interest rules and guidelines, and by explicitly informing their panelists about such rules and guidelines. Doing so not only makes the peer-review system fair and objective but also more effective by preventing favoritism and cronyism, and avoiding awards for undeserving proposals. As mentioned in Section 6.8, program directors have a particular responsibility for making peer review process work fairly and effectively.

6.9.3   Mistake 3: Lack of calibration in funding levels

Funding agencies support research projects under a multitude of programs. Each of these programs is given an operating budget and programs directors are given almost complete autonomy to decide the amount of funding for each proposal they choose to support. As most researchers are willing to settle for any amount of funding, this leads to arbitrary levels of funding across programs and sometimes even within the same program. As a result, some investigators end up receiving one graduate student support while others get two, and some receive one-month summer support while others get two-months of summer support, and so forth. There should be room for program directors to negotiate budgets with investigators but such negotiations should not degenerate to unworkable budgets or irregular amounts of funds for similar research projects. To alleviate these issues, funding agencies can develop guidelines to help program directors negotiate project budgets effectively with investigators and without compromising the integrity of projects. Differences in funding amounts due to salary and other cost factors across institutions of investigators should be kept to a minimum to motivate researchers to use their funds as effectively as possible.

6.9.4 Mistake 4: Excessive project reporting requirements

As stated earlier, funding agencies require project investigators to report their findings at regular intervals of time. It is a good idea to probe the progress of research projects, but this can easily be overdone to cause investigators to spend much of their valuable research time on reporting functions. The reporting periods can be adjusted depending on the expectations of a funding agency. However, it is counterproductive to ask investigators how many new papers they have been able to publish every two months or how many doctoral students they supervised to completion every six months. The annual project reporting process used by NSF allows investigators to submit their results in a timely fashion without too much overhead, and may serve as an example to other funding agencies that require more frequent reporting.

6.9.5 Mistake 5: Transparency and accountability issues

It is important for government funding agencies to disclose their funding information in a timely manner to the public. For example, NSF periodically reports on its merit review process to the National Science Board.* This gives NSF an opportunity to document its actions for the public and to see if there is any room for improvement. NSF also uses a committee of visitors process to evaluate the performance of its divisions and programs by outside experts. In the most recent merit review document, NSF reports that 99 such committees were invited to evaluate various NSF divisions and programs between 2004 and 2007. These visits point out a number of issues pertaining to panel operations and decisions, and suggest ways to fix problems. It will be helpful if NSF also reports what actions it has taken to address the concerns raised by these committees of visitors. Other funding agencies may benefit from such external evaluations and institute similar reporting and external review processes to make their operations more transparent and accountable to their constituencies.

6.10 Summary

This chapter presented the funding agency perspective of research and described a funding model to characterize the proposal processing and evaluation, and funding decision processes in funding agencies. It discussed the operational structure of funding agencies, and stressed the important role that program directors and review panels can play in making funding decisions more than guesswork. The chapter also described

---

* National Science Board, p. 31.

specific methods for classifying proposals into panels, assigning reviewers to panels, and ranking proposals. Finally, the chapter reviewed some of the common mistakes funding agencies can avoid to make their funding decisions more effective.

6.11   Bibliographical notes

The material on proposal clustering and panel assignment problems has been the subject of a research project funded by the Scientific and Technological Research Council of Turkey under Grant No. 109M149. Some of the results obtained in this research project were originally reported in the following publications.

1. Oruc, A. Y., and Atmaca, A. 2009. *Asymptotically optimal assignments in ordinal evaluations of proposals.* arXiv:0908.3233v1.
2. Oruc, A. Y., and Atmaca, A. 2009. *On ordinal covering of proposals using balanced incomplete block designs.* arXiv:0909.3533v1.
3. Yıldırım, A., Aykanat, C., and Oruc, A. Y. 2009. *A comprehensive electronic proposal evaluation and selection system.* TÜBİTAK Project. Grant No. 109M149.
4. Oruc, A. Y. 2008. *A combinatorial analysis of ordinal ranking problems.* Unpublished manuscript.

6.12   Questions

6.1   Do you find it difficult to identify the program that funds your research in your funding agency? Do you think that you can submit your proposals to more than one program? If so, how do you decide to which program to submit your proposals?

6.2   Which do you prefer for your proposals: mail reviews or panel reviews? Why?

6.3   How do you view your role as a reviewer for your fellow researchers in the field? Does knowing them through professional contacts, for example, serving on program committees and editorial boards together, affect your reviews? How do you feel about your proposals being reviewed by someone you professionally know? Would you prefer that proposals be reviewed using blind reviews? Why or why not?

6.4   Would you like to work on research problems that significantly depart from the mainstream of your research field or those that remain within the mainstream?

6.5   Does the idea of fellow researchers serving as program directors at your funding agency appeal to you? Do you feel that they can be impartial and adhere to high standards of conflict of interest ethics?

6.6   If you could serve as a program director at a funding agency, how would you select reviewers? Would you rely on your

professional contacts alone or solicit suggestions from researchers you do not know professionally. What criteria would you use
to select your reviewers and panelists?

6.7   Show that no cut of six edges can partition the weighted proposal graph in Figure 6.2 into two subgraphs of six vertices with
a combined affinity that is less than 1.4.

6.8   Partition the weighted proposal graph in Figure 6.2 into three
subgraphs graphs of four proposals so that the total affinity of
cut edges is minimized.

6.9   Partition the weighted proposal graph in Figure 6.4 into three
graphs of four proposals so that the standard deviation of the
affinities of proposals is minimized.

6.10   For the reviewer-proposal pairing matrix shown next, compute
the maximum number of pairs of proposals assuming that each
reviewer has a capacity of five proposals.

$$X = \begin{array}{c} \\ r_1 \\ r_2 \\ r_3 \\ r_4 \end{array} \begin{array}{cccccccc} p_1 & p_2 & p_3 & p_4 & p_5 & p_6 & p_7 & p_8 \\ \begin{bmatrix} 1 & 1 & 0 & 1 & 1 & 1 & 1 & 0 \\ 0 & 1 & 1 & 1 & 0 & 1 & 1 & 1 \\ 1 & 0 & 1 & 1 & 0 & 0 & 1 & 1 \\ 1 & 1 & 1 & 0 & 1 & 1 & 1 & 0 \end{bmatrix} \end{array}$$

6.11   If all n proposals on a panel must be compared on a pairwise
basis and if each panelist can review at most $n/6$ proposals, how
many panelists are necessary? How many are sufficient?

6.12   Give an assignment that covers all pairs of 24 proposals using 12
reviewers each with capacity 8.

6.13   Give an assignment that covers all pairs of n proposals using the
number of reviewers given in Remark 6.7.

6.14   For the following set ranking of proposals by three reviewers,
determine if a ranking (partial ordering) of proposals exists.

$P = \{p_1, p_2, p_3, p_4, p_5, p_6, p_7, p_8, p_9\}$ and $R = \{r_1, r_2, r_3\}$ with the rankings
$\preccurlyeq r_1 = \{(p_1 \preccurlyeq p_2), (p_2 \preccurlyeq p_3), (p_1 \preccurlyeq p_3), (p_1 \preccurlyeq p_8), (p_8 \preccurlyeq p_9), (p_1 \preccurlyeq p_7)\}$
$\preccurlyeq r_2 = \{(p_6 \preccurlyeq p_2), (p_6 \preccurlyeq p_5), (p_5 \preccurlyeq p_4), (p_6 \preccurlyeq p_4), (p_6 \preccurlyeq p_7), (p_9 \preccurlyeq p_1)\}$
$\preccurlyeq r_3 = \{(p_3 \preccurlyeq p_4), (p_4 \preccurlyeq p_8), (p_3 \preccurlyeq p_8), (p_9 \preccurlyeq p_7), (p_4 \preccurlyeq p_7), (p_6 \preccurlyeq p_9)\}$

6.15   Can you think of any mistakes that funding agencies make other
than those described in Section 6.9?

Index